Petroleum Politics and the Texas Railroad Commission

NUMBER TWELVE
The Elma Dill Russell Spencer Foundation Series

Petroleum Politics and the Texas Railroad Commission

by David F. Prindle

University of Texas Press, Austin

For my parents, Elliott and Vivian Prindle, in partial compensation

Printed in the United States of America
Second Paperback Printing, 1985

Library of Congress Cataloging in Publication Data

Prindle, David F. (David Forrest), 1948–
Petroleum politics and the Texas Railroad Commission.
(The Elma Dill Russell Spencer Foundation series; no. 12)
Includes bibliographical references and index.
1. Petroleum industry and trade—Texas. 2. Energy development—Texas. 3. Petroleum law and legislation—Texas. 4. Energy development—Law and legislation—Texas. 5. Railroad Commission of Texas. I. Title. II. Series: Elma Dill Russell Spencer Foundation series; no. 12.
HD9567.T3P74 353.97640087'5'06 81-7535
ISBN 0-292-76489-8 pbk. AACR2

Contents

Preface

Much of the background information, and some of the specific facts on which this study is based, came from interviews with people familiar with the Railroad Commission of Texas or some relevant aspect of the petroleum industry. Early in the research, it became obvious that few individuals would discuss the subject freely unless they were promised anonymity. I therefore adopted a policy of assuring my respondents that they would not be identified. For this reason, many general statements and even a few direct quotations in this book are not attributed. This is of course unfortunate, but the alternative—noncooperation—is worse. I have attempted to provide as many citations to published material as possible to back up my unattributed assertions.

Of the ninety-three men I interviewed, however, one requested that he be publicly identified. Fair enough: I am very grateful to Johnny Mitchell of Mitchell Energy Corporation for talking with me and giving me access to his private files. Mr. Mitchell's help was invaluable in the preparation of chapters 4 and 5 of this book.

Aside from Mr. Mitchell, I interviewed six former or sitting Commissioners, engineers, attorneys, and others employed or formerly employed by the Commission, a handful of executives of major companies, over a dozen independent oil producers, journalists, trade association executives, consumer advocates, municipal officials, and fellow researchers. Some of these categories tend to overlap. I spoke with several of these people for as little as ten minutes by phone; some I visited repeatedly over the course of two years. I have several hundred hours of tape recordings of the lengthier interviews, for those qualified scholars who wish to check my sources.

I am under no illusion that I could have completed this work without the cooperation of these ninety-two people. Although I will not name them, I must thank them publicly for their generosity. They have made this book possible.

Interviews, however, supplied only part of the information on which my study is based. I also drew on newspapers and other published sources, archival material, court decisions, and private collections of correspondence. Because of space restrictions, this book does not contain a bibliography. I hope that my notes are sufficient to indicate my noninterview sources.

Scholars of Texas history may be puzzled about the lack of citations to the Railroad Commission minutes in the Texas State Archives. There is no mystery: I perused forty years of those minutes and found nothing useful. They consist mainly of lists of transportation licenses and contain no relevant reports of discussions among the Commissioners or testimony at hearings.

Although this study was conceived and carried through entirely by myself, both the Center for Energy Studies and the University Research Institute at the University of Texas provided me with financial support. I am particularly grateful to Sally Cook, Lorna Aldrich, and Herb Woodson of the CES for their encouragement. In view of the controversial nature of the subject of this study, I must point out that no one associated with either the CES or the URI ever attempted to influence or deflect the direction of my inquiries or the formulation of my interpretations. They played the role of disinterested patrons only.

A number of undergraduate research assistants, by doing much of the detailed work, made this project bearable for me. Sandra Nicolas, James Burshtyn, Clea Efthimiadis, Kamala Collins, Kerry Kilburn, Sylvia Mandel, and Mary O'Connell were reliable, intelligent helpers and friends.

A petroleum engineer, long employed by the Railroad Commission, read eight chapters of the first draft of the manuscript. He prefers not to be named. Professor William O. Huie of the University of Texas School of Law and Professor Stephen L. McDonald of the University of Texas Economics Department each read six chapters. All three of these men provided me with thoughtful and constructive criticism. Part of any merit this book possesses is due to their willingness to help. As with any similar undertaking, however, there were inevitably some irreconcilable differences of opinion. In other words, I did not always take their advice. They cannot be blamed for whatever errors of fact or interpretation remain.

Some of my colleagues in the Government Department read one or several chapters and made many helpful comments. The brevity with which I must thank them is not commensurate with the service they have done me. They are Jeff Millstone, Gordon Bennett, William Galston, Isabel Marcus, Richard Kraemer, Janice May, Karl Schmitt, and Alfred Watkins.

Various other people were helpful far beyond the requirements of their jobs or even the demands of good manners. I cannot thank each in full, but every one contributed in some essential way to the completion of this project: Warren Anderson, Charles Bowlin, Felizia McDaniel, George Shipley, Richard Murray, Joyce Nettles, Tedd Cohen, David Patterson, William A. Rennie, Mary Ann Tetrault, and Paul Sweeney.

Kelly McWhirter typed the final version of the manuscript. Not only did she finish it on time, which I expected, but she cheerfully put up with my constant interposition of last-minute changes, which I didn't.

Part 1

Introduction

1. The Railroad Commission of Texas

To govern is to choose.—Political proverb

This is a book about choices. For fifty years, choices have been made in Austin, Texas, that have vitally affected United States oil policy. How much petroleum is produced, how much is discovered, its price, the number of people who do the producing, and their relationship to one another are just some of the aspects of the oil and gas industry that have been shaped by the choices made by the sixteen men who have served as Texas Railroad Commissioners since 1930. Because these choices have involved the supply and price of a vital resource, and because they have resulted in the distribution of billions of dollars to private individuals, they have been the subject of intense political struggle. But, because they have been made in a city far from the beaten paths of national policy–oriented reporters and social scientists, the political process that produced them has generally been ignored.

This neglect of the Railroad Commission is understandable but unfortunate, for, since the mid 1930s, it has been one of the most important regulatory institutions in the United States. Such federal agencies as the Interstate Commerce Commission and the Federal Communications Commission exert great power over some part of the national economy. But the Railroad Commission, though it is only a state agency, deals with a commodity that underlies virtually every aspect of modern industrial life. It is not just that petroleum fuels transportation, drives machinery, and heats buildings. The events of the 1970s highlighted the fact that petroleum-based products are used to feed livestock, spin synthetic fabrics, and manufacture phonograph records. In fact, there is oil in almost every useful or entertaining product in the country.

A glance at the *Fortune 500* list of United States corporations in the late 1970s confirms the fact that eight of the fifteen largest corporations in the nation are oil companies.[1] Two of these corporations, Gulf and Texaco, got their start in the Lone Star State, and several of the others, including Exxon, Mobil, and Shell, have large holdings there. Regulations propounded by the Railroad Commission, therefore, have directed the activities of the industrial giants of the American economy.

This is true because Texas has long been the pivotal petroleum-producing state. From the early 1930s until the 1970s, the state produced from 35 to 45 percent of the national total of oil each year. Alaska now competes with Texas for the title of number one oil state, as each is responsible for about 30 percent of the United States total.[2] And of course the increasing national reliance on foreign supply has further diminished Texas' importance as a source of oil. But, as oil has fallen, natural gas has risen in importance. Because Texas produces over a third of the natural gas in the country, and because the Railroad Commission regulates gas as well as oil, the Commission will be nearly as important in the 1980s as it was in the preceding decades.[3]

PETROLEUM AND DEMOCRACY

If the Railroad Commission were noteworthy only for its relative obscurity despite such influence, it would still be well worth studying. But the Commission possesses an additional attribute that recommends it to students of politics: Commissioners are elected. The occupants of federal regulatory agencies are appointed. No state regulatory commission, elective or appointive, approaches the national importance of the Railroad Commission. The Commission is consequently the only nationally important regulatory agency whose members must periodically answer to the voters.

Scholars who have expended a great deal of effort studying patterns of influence in national regulatory commissions have developed a fairly consistent set of findings. In general, after a short post-creation period in which it assumes an adversary stance in relation to its particular industry, the typical agency becomes coopted by that industry. Instead of being a means by which the regulators oversee an industry in the public interests, the agency becomes a means by which the industry regulates itself in its own interests. This happens at least partly because commissioners are appointed, often for long terms, and are therefore not subject to the voters. Many studies have described the process by which regulators become captured by industry information, industry argument, and industry socializing.[4]

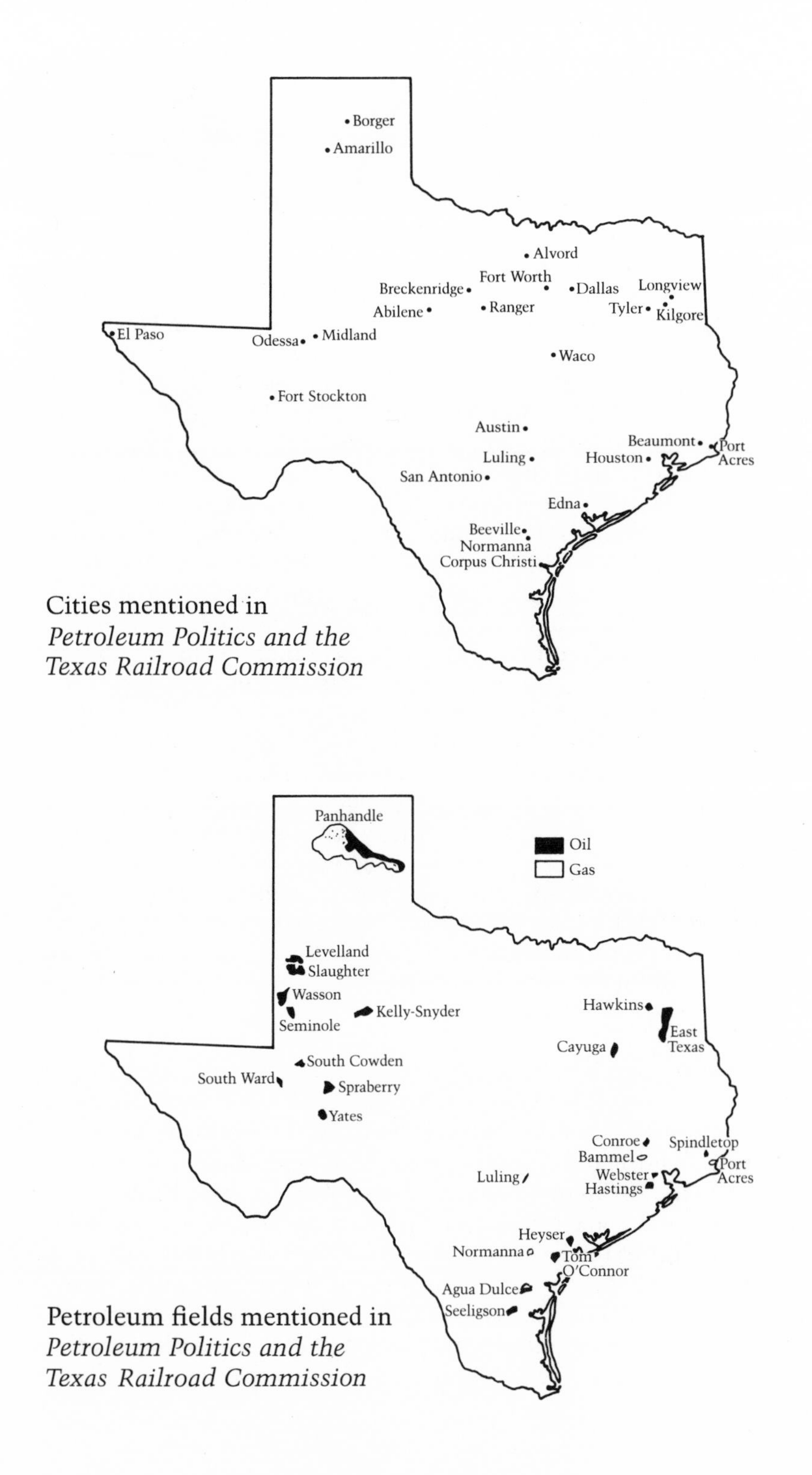

Cities mentioned in *Petroleum Politics and the Texas Railroad Commission*

Petroleum fields mentioned in *Petroleum Politics and the Texas Railroad Commission*

This process has gone so far and has become so well known that there is now a strong movement to deregulate such industries as plane and truck transportation, on the theory that the public interests can be better served by competition than by regulation.

But what of an elective commission? In Texas, there is an election for at least one Railroad Commission seat every two years. When the Commission was created, under Governor Jim Hogg's administration in 1891, it was intended to keep watch on the railroads—hence its name. Hogg wanted an appointive agency, for he thought that the railroads might buy elections but could never buy him.[5] The Commission's importance now rests on its responsibility for a different industry, but the questions for scholars of democratic practice remain: are Commissioners, like other regulators, coopted apologists for the petroleum industry? Or, because they are elected, are they responsive to the voters' desires?

Part of this book will be devoted to demonstrating that, because Commissioners have historically perceived Texas' interests to be advanced by a healthy oil industry, they have indeed adopted the industry's perspectives and endorsed its goals in many issues. But the analysis will continue to show that the structure of the state electoral process and the Commissioners' own views on state needs have led them to endorse one segment of the industry to a far greater extent than another.

Specifically, a major theme of this book will be that Commissioners have consistently chosen to side with the independent producers and small landowners, as opposed to the major integrated companies, in their policy making. Several chapters will explore the circumstances that have led to this political outcome as well as its consequences for the nation.

WHAT THE COMMISSION DOES

The regulations that have made the Railroad Commission the pivot of national petroleum policy fall into three broad categories.

First, the Commission is a conservation agency. Oil does not flow to a well by itself. It must be pushed, usually either by the expansive power of the natural gas caught in the rock layers with it or by water pressure. If the oil is produced too fast, these natural drives may dissipate, and much of the oil may be trapped underground, possibly forever. Since the mid 1930s, the Commission has set the maximum permissible rate of production for both oil and gas fields in the state and, by so doing, has prevented the waste of these natural resources.

Additionally, petroleum production can be hazardous to humans and, in a larger sense, to the natural environment. Leaking gas can cause explosions, uncontrolled oil wells can foul streams and beaches, improper drilling can pollute underground water reserves, abandoned wells can trap children, and so on. To ensure safety and physical conservation, Commissioners set rules of procedure for drilling wells, and Commission field inspectors oversee the drilling. Moreover, the Commission regulates pipelines to ensure that they are operated safely. This first aspect of the Commission's responsibility is plainly important, but it is not controversial and rarely excites political disagreements.

Second, however, the Railroad Commission has historically been responsible for prorating oil production to market demand. From the early 1930s to the early 1970s, there was a worldwide oil surplus. During this period the Commission suppressed each Texas field's output well below the level dictated by a concern for mere physical conservation, matching the state's total production to the expected market demand. Each well in each field was assigned an allowable, which defined the proportion of the theoretical maximum that the well could produce each month. That is, the Commission permitted only enough oil to be drawn out of the ground to satisfy the current demand at the current price.

Market-demand prorationing had three great consequences. It encouraged exploration beyond what was necessary to satisfy short-run demand, thus maintaining a surplus production capacity. As a result, when there was a sudden increase in the demand for oil, as during World War Two and the Middle East crisis of 1967, production could be quickly expanded to prevent a shortage. Moreover, prorationing restrained the supply of oil, so that too much production did not make prices fall. Finally, it maintained a host of high-cost, independent producers who would not have been able to survive if the forces of supply and demand had been allowed to operate unfettered. By so doing, it greatly improved the economy of Texas.

Because demand for oil became so high in the 1970s, the Railroad Commission does not prorate that substance to market demand anymore but allows most wells to produce up to a level that will not damage the reservoir. However, it continued to prorate statewide gas production to a sort of market demand through the 1970s, with consequences to be explored in chapter 5.

The practice of market-demand prorationing has been the subject of angry controversy. From the 1930s to the present day, Commission critics have charged that it amounts to state-sanctioned price-fixing. Prorationing was attacked first by independent producers and landowners with evasion, lawsuits, and violence, later by economists and consumer advocates with monographs, exposés, and theoretical treatises. This conflict, naturally an important part of the subject matter of this book, will be examined in chapters 2 through 6.

Third, Railroad Commissioners have at all times striven to protect the correlative rights of every producer and royalty owner in the state. The phrase "correlative rights" is both vague and changeable, but it refers at bottom to two physical facts and one economic circumstance. Physically, oil and gas are invisible underground, and both move, or flow, across property lines. Economically, they are valuable. In combination, these conditions guarantee that, in societies where private ownership of land is the rule, there will be frequent conflict over the right to produce and market petroleum.

Because oil and gas are valuable, anyone with a connection to ownership of underground petroleum wants to make sure that the fullest possible reward from that ownership is realized. Because petroleum is invisible and flowing, however, the determination of ownership, and thus the determination of rewards, is a process consisting in large measure of speculation. It has been the unenviable task of the Commissioners to attempt to referee the production of oil and gas in Texas so that no one feels cheated. As might be expected when fortunes are involved, this has been a dangerous and highly politicized task. In one case—the slant-well episode—protection of correlative rights was largely a matter of first identifying the producers who were stealing their neighbors' oil and then overcoming these producers' considerable power to delay investigation. But, in most cases, protection of such rights has meant dealing with much less tangible problems.

For example, the Railroad Commission has attempted to ensure that buyers of oil and gas will purchase from producers in a manner that will guarantee every operator a market, instead of allowing them to buy in some discriminatory but more profitable manner. Similarly, the Commission has tried to see to it that transporters (chiefly pipeline companies) do not refuse to carry petroleum. Since the 1930s, the Commission has been successful in thus ensuring equitable treatment to oil producers, as explained in chapters 3 and 4. Because of the different technical and economic situation of natural gas, however, the Commission has never quite brought the production and transportation of that fuel under control. As explained in chapter 5, Commissioners continue to struggle with the regulation of gas.

In addition, having the responsibility for adjusting correlative rights has forced the Commission to deal with the problem of allocating production quotas among producers who possess tracts of varying potential. The power to prorate involves an obligation to set some rules by which production can be allocated. Naturally, all operators want to be permitted to develop their tracts in an advantageous manner. If the amount of land they control is small, however, or if their wells are positioned close to property lines, profitable production might result only from their draining the oil and gas from under their neighbors' land.

As explained in chapters 3 and 4, for many years it was the policy of the Commission to discriminate in favor of the little guy in setting its spacing rules and production allowables, thus encouraging production on even very small tracts. This pattern changed in the early 1960s, because of hostile court decisions and personnel turnover on the Commission. Allocation and spacing rules are now much less discriminatory.

In dealing with each of these problems, the Commission has been the arena of intense political maneuvering, and the choices the Commissioners have made have determined economic outcomes in both Texas and the rest of the country.

STUDYING THE COMMISSION

In order to discover what the Commissioners have done and why, it will be necessary to study their major policies systematically. This book will therefore be partly a history, for the Commission's importance begins in 1930 and continues to the present.

Students of public policy are not agreed on the best method of studying decision making by politicians. There are many suggested approaches to the study of the decision-making process in government but few satisfactory studies of actual decisions. Even the best treatments emphasize both the difficulty of discovering what those in power were thinking and the ambiguous nature of political choice.[6] One text on the subject offers eighteen different models of policy making, a number almost guaranteed to encourage intellectual paralysis rather than informative research.[7]

To study the important policy decisions of the Railroad Commission with a semblance of order, and to help focus on both the process of influencing political choice and its consequences, a methodology will be adopted which combines the approaches of several other studies.[8] Because the major Railroad Commission policies involve not one clear decision but a series of greater or lesser adjustments to circumstances, what will be analyzed are not so much decisions as policy-making episodes. Commission choices on most important topics evolved, sometimes incrementally, sometimes radically, over decades. Over a dozen individuals often participated in hundreds of decisions, all of which combined to make a policy.

The purpose of examining policy-making episodes is to uncover both trends in regulatory decisions and the political environment that could produce such trends. It would be an overwhelming task to study every policy that has emerged from the Commission for five decades. Some criteria must be applied to help choose only the most educational episodes. Those that were picked for investigation were the ones that satisfied the following criteria:

1. Decisions as opposed to nondecisions. No doubt, some topics that the Commissioners neglected are important, but trying to study them would make for a short and uninformative book.

2. Conflict. There must have been considerable disagreement, either between the Commissioners themselves, and/or between segments of the industry, and/or between the industry and the larger public, or, preferably, all three. Political conflict often inspires people to record their positions publicly, in speeches, pamphlets, letters, and so on. Such traces allow us to interpret the coalitions that formed around the issues with which Commissioners dealt.

3. Consequences. Railroad Commission policy making is a classic case of authoritative decisions determining who gets what: who gets to explore, drill, produce, transport, and so on.[9] To be included in this book, a policy decision must have had significant consequences for the price and availability of petroleum and for the balance of prosperity within the industry.

When these criteria were applied to the Railroad Commission's history, seven policy-making episodes seemed indispensable:

1. Establishing the right to prorate oil, 1930–1935. This was partly an external struggle, in which the Commissions' authority was gradually acknowledged by the courts, the federal and state governments, and the industry, and partly an internal struggle within the industry.

2 and 4. Well spacing and allocation of production within fields, 1930–1965. Ideally, a very few wells can drain most petro-

leum reservoirs. But it is often in the interests of individual landowners to drill many wells, thus making investments which are inefficient from the perspective of society at large. Commissioners have been responsible both for making rules governing the spacing of wells and for allocating production among those wells according to some defensible criteria. This particular episode naturally separates into two parts, one before and one after 1946.

3. Gas flaring, 1930–1949. Prior to the 1950s, oil was valuable, and natural gas was not. Producers consequently viewed gas as a useless by-product and burned it in huge quantities. The Commission, in the name of conservation, finally forced them to save and store this gas instead of wasting it.

5. Slant wells, 1946–1963. Oil migrates. Suppose drillers deflect their well bores to follow the moving oil and end up under someone else's property? In attempting to prevent this, Commissioners became caught in a regulatory quagmire and, in 1962, a great scandal.

6. Gas prorationing, 1930–1980. Largely because gas was not valuable in the 1930s, the authority to prorate its production was not established, as it was for oil. As gas became more important in succeeding decades, the Commission attempted, with only partial success, to regulate its production.

7. The Lo-Vaca problem, 1972–1979. A gas utility, supplying a third of the population of Texas, became unable to meet its contracts. The Commission was caught in cross fire between consumers, the industry, and the energy crisis.

Once the subject of the study had been chosen, there remained a problem about what method of approach would be most helpful. It is important to discover not only the political atmosphere on the Railroad Commission itself when policies were adopted but also the nature of the conflict at issue, as it seemed to people in the industry at the time. It will become clear as the narration proceeds that it is impossible to understand the choices of the Commissioners unless the divisions within the industry are first comprehended; it is impossible to understand those divisions unless something is known of the economic structure of the industry; and it is impossible to understand that structure without some grasp of the changing technology of oil and gas production.

Consequently, in common with other explorations of decision making, the methodology of the present study will focus on who actually made decisions, against what opposition, or who prevented decisions from being made. Further, this methodology will also incorporate a desire to discover what individuals were thinking when they made, or refrained from making, decisions as well as what social groups won and lost because of the particular policies that were adopted. Finally, it will be necessary to incorporate discussions of the physical situation and composition of hydrocarbons, and their relationship to economic circumstances, into the analysis of political interests. This will make for a complex discussion, but anything less would distort the political reality.

For each of the seven major problems faced by the Railroad Commission, these questions will be asked of the historical record:

1. What was the technological problem? That is, how did geological, chemical, economic, legal, and historical circumstances combine to force conflict on the industry?

2. In what terms was the problem perceived at the time? How did the participants define the available alternatives, and how did they understand their own role in contributing to the final choice?

3. Who made a decision that prevailed? What was the opposition? Why did that particular position succeed rather than another and why at that time rather than another? Or, conversely, who prevented a decision from being made?

4. What were the consequences of the decision? What groups or individuals won economic rewards or political power because of the decision, and what groups lost? Did the decision alter the economic or political context, so that it affected future decisions?

MAJORS AND INDEPENDENTS

In the course of answering these questions, it will become clear that two industry perspectives have repeatedly clashed in the Commission's councils. As with any industry, petroleum has many internal conflicts, some of which have spilled over into the public arena. Unlike many other industries, however, petroleum is more or less permanently divided into two segments that traditionally have many different economic, and therefore political, interests. Some of the political episodes recounted in this book are largely discussions of this abiding conflict between majors and independents. A short discussion of each group will help set the context for the analysis to follow.

The Major Integrated Companies

There are many methods of distinguishing the majors from the rest of the industry.[10] The most common criterion, and the one used in this book, defines any oil corporation that integrates all four of the basic activities of the industry—producing, transporting, refining, and selling at retail—as a major company.

Almost by definition, majors are enormous industrial agglomerations. They include some of the largest and most profitable companies in the world. This fact has affected their political interests. As established and bureaucratized organizations, with far-flung investments and diversified activities, majors value a stable, predictable industry. They avoid risks and strive to impose order and control on industry development. They typically provide a cosmopolitan, long-run perspective on industry controversies.

The Independents

The best way to define this group is by decreeing that they are what is left over after the majors have been subtracted from the industry. Of the four main activities of the industry, independents perform no more than three. They are not integrated.

But this definition does not begin to suggest the variety of independent petroleum companies. A few independents produce, transport, and refine and are so large that their economic perspective is identical to that of the majors. Some independents only transport oil and gas (mainly through pipelines) and therefore rarely participate in the battles that rage over producing interests. Whenever an "independent's perspective" is mentioned in this book, it must inevitably be something of a distortion, because the diversity of interests among independents makes any effort to aggregate their viewpoints misleading.

Nevertheless, there is a balance of activity among independents that makes generalization about their interests both possible and fair. The great majority of independents are small oil and/or gas producers only. The 1978 *Oil Directory of Texas* lists almost six thousand of these independents, better than two-thirds of them producing less than 1,000 barrels of oil and 10 million cubic feet of gas in a typical month. By way of contrast, Texaco produced 4.75 million barrels of oil and 17.75 billion cubic feet of gas in a month.[11] Very few of these small independents operate outside of Texas; most confine themselves to a general area or even to a single field. Opinion among the independents is consequently predominantly that of small business, for that is their accurate perception of themselves.

In the history of the oil industry, the independents' concentration on production, their generally small size, and their lack of geographic dispersion have had a decisive effect on their perspective on regulatory policy. Independents are the risk takers in the industry (more on this in chapter 9). They do not normally place great value on stability or corporate order. In controversies within the industry, their viewpoint has usually been short-run and parochial.

Because independents are often important actors in the local economies of Texas, their perspective is often shared by the small business people on the fringes of the industry. Small landowners (that is, present or potential royalty owners), drillers, small refinery operators, saltwater haulers, and the other persons who derive a livelihood from the spin-off industries of the petroleum fields frequently share the independents' interests. In the course of this book, "independents" or "independents and small landowners" will often be used to mean the coalition of locally oriented small businesses that share the independents' perspective on regulatory policy.

CONCLUSION AND PROLEGOMENA

This book contains an analysis and an evaluation of half a century of public policy choices by the Texas Railroad Commission. It will focus on the important decisions made by the Commissioners, on the political processes that led to those decisions, and on the consequences of those decisions for Texas and the United States.

There will be many detailed conclusions in this book, as well as many more detailed qualifications. But some findings will stand out amid the ambiguities and mysteries: for fifty years, the Railroad Commission has been an agency devoted to advancing the economy of Texas, sometimes at the expense of the rest of the country. As part of this effort, Commissioners have traditionally attempted to protect the petroleum industry, in particular its independent side, from the hazards of unrestrained competition. In addition, however, the Railroad Commission has gone beyond the limits of simple protection of independent oil companies and has favored smaller producers and royalty owners at the expense of larger producers and owners. As a consequence, it has favored producers in general over consumers in general and Texas over the nonproducing states. In the early years, this favoritism was pronounced, but it has been considerably mitigated in the intervening decades. Nevertheless, the tradition of favoring the little guy still exists and affects Commission policy choices.

That the Commissioners have done this does not necessarily mean that their actions have been contrary to the national interests.

Whether the policies pursued by the Commission should be praised or condemned is a complex question to be discussed in chapters 6 and 9.

Commissioners have followed a consistent pattern of choice because of the political context in which they have operated. As will become clear in chapters 7 and 8, the influence of the independents has been magnified by peculiarities in the state's economic and political structure. As a result, despite the fact that Commissioners are elected, they are most responsive to a relatively small part of their theoretical constituency. An examination of this system, of its origins, structure, and consequences, is the task of the rest of this book.

Part 2

Policy

2. Foundations, 1930–1935

It must be considered that there is nothing more difficult to carry out, nor more doubtful of success, nor more dangerous to handle, than to initiate a new order of things.—Machiavelli, *The Prince*

In the late nineteenth century, Texas was a hotbed of agrarian radicalism. As a people whose economy was based almost completely on cattle and cotton,[1] Texans were particularly sensitive to the price of commodity transportation and thus felt themselves at the mercy of the railroads. In the economic dislocations that followed the Civil War, many farmers came to feel that the practices of the railroads were adding to their hard times. Resentment of the railroads, which was an important part of the impulse that moved Texans to organize the Farmers' Alliance in 1875 and 1879, was also a strong force behind the rise of the Populist movement in the 1890s.[2]

Riding the crest of the antirailroad wave in Texas was Jim Hogg. Hogg, who acquired his political reputation by winning cases against railroads as attorney general,[3] ran for governor in 1890 promising to establish an agency to oversee the state's transportation industry. There was no doubt about the popularity of such a proposal, for Hogg won by a large margin.[4] The problem came in implementing the plan. Hogg wanted an appointive commission, for he felt that railroad money would be too powerful in elections, making Commissioners subject to capture by the industry. The railroads, perhaps agreeing with this assessment, lobbied hard for an elective agency. Hogg succeeded in getting both a strong Commission and the power of appointment, but he lost the latter through a political blunder. Running for governor, he had asked for and received the support of the Farmers' Alliance, and its members had expected that, once in office, he would appoint at least one farmer to the Commission. For unaccountable reasons, he failed to do so. Farm organizations immediately launched a campaign to create an elective agency, and the legislature responded three years later by modifying the law.[5] Commissioners henceforth would have to answer for their decisions to the voters every six years.

The actions of the Commissioners in dealing with railroads in the subsequent ninety years are of only limited interest. But in 1901 an event occurred that would ultimately make the Commission a vital cog in national government. On January 10, Captain Anthony Lucas, an Austrian engineer, brought in a spectacular oil gusher at Spindletop, just south of Beaumont. This was not the first oil well in Texas, but it was so prolific—capable initially of producing one hundred thousand barrels a day—that it quickly became the basis for a statewide preoccupation with petroleum.[6] Engineers, drillers, wildcatters, promoters, organizers, roughnecks, and swindlers swarmed to Texas from all over the country, and in the next thirty years they and homegrown producers created an important industry. Oil was discovered along the Gulf coast, in the eastern piney woods, on the inland prairies, and in the deserts of the west. Ranger, Borger, Breckenridge, and several other small cities became legendary boomtowns; penniless farmers turned into wealthy royalty owners; and corporations transformed the agricultural economy. By 1929, Texas had produced over two billion barrels of oil and was the number one producing province in the world.[7]

In a roundabout way, the Railroad Commission was given the authority to regulate a portion of this industry. At the beginning of the twentieth century, Texas' antimonopoly laws, a holdover from its Populist phase, forbade any corporation to engage in more than one aspect of the petroleum business. That is, of the operations of producing, transporting, refining, and retailing, a corporation could engage in only one. Since this integration of operations is the very essence of a major oil company, the Texas laws were, to say the least, inconvenient for the non-Texas majors, and they perpetually campaigned to have them repealed. Independents, that is, nonintegrated companies, opposed a change in the laws and managed to stalemate the majors for over a decade. In 1917, however, a compromise was reached: the antiintegration laws were repealed, but the state assumed the right to regulate pipelines. Since pipelines carried petroleum and were therefore a means of transportation, the Railroad Commission was given the authority to oversee their operations.[8] Two years later, when the legislature passed a statute prohibiting the waste of natural resources, especially oil, the Commission was naturally expected to enforce it.[9]

During the 1920s, the Commission promulgated rules to prevent the waste of oil, such as Rule 37, the well-spacing rule, but in general it was not an agency known for its vigorous activity. The oil industry was healthy, and both the industry and the Commission were inclined to leave well enough alone.[10] In 1930, however, with the Great Depression clamping down, the industry experienced an extraordinary stroke of misfortune, which created an entirely new set of expectations for the Commission.

POLICY-MAKING EPISODE NUMBER ONE
The Institutionalization of Oil Prorationing

THE GREAT FIELD

On October 3, 1930, a promoter named Columbus M. ("Dad") Joiner brought in a well on the Daisy Bradford farm three miles south of the hamlet of Kilgore, roughly between Longview and Tyler in Northeast Texas, thus discovering what was to become known as the East Texas field. The oil pool discovered by Joiner was unusual in three respects, all of which combined to throw the world of oil into chaos.

First, East Texas was an enormous reservoir, measuring forty-five miles north to south and from five to twelve miles east to west. It covered over 140,000 acres and contained about 5.5 billion barrels of oil.[11] This meant that it was the largest oil pool that had been discovered in the world—it contained perhaps a third as much as all the oil produced in the United States up to that time.[12]

Second, the East Texas field was the product of a kind of geological circumstance with which oilmen were not then familiar. Most of the oil that had been discovered up to 1930 was congregated in rock formations—such as faults or folds—that could be detected easily by surface observation or seismic equipment. The East Texas field, however, was the product of a stratigraphic trap, in which a layer of rock named the Woodbine sandstone simply ended in a pinch-out (a sideways, elongated V) about 3,600 feet below Kilgore. Because this stratigraphic trap was not associated with any disturbance of the rock layers, it could not be detected by familiar techniques. For this reason, the field was discovered not by a conventional geologist or seismic physicist but by a wildcatter.[13]

The eccentric geological situation of the East Texas field was responsible for its third unusual quality: the major companies had failed to acquire leases on most of the producing surface acreage. Although several Humble geologists had suggested, and for the right reasons, that the area contained oil, neither Humble (now Exxon) nor any other major company had set about buying drilling rights in

East Texas with enthusiasm.[14] As a result, a year after Joiner's discovery, the majors owned less than 20 percent of the wells and proven acreage.[15] East Texas became the promised land of independent producers and small royalty owners, a fact that was to determine the politics of oil in Texas for the next fifty years.

OVERPRODUCTION AND CONSERVATION

The oil industry is not like other businesses. In most other trades, supply and demand come into relative balance more or less automatically. Manufacturers of shoes, for example, estimate how many units of their product will sell in the coming year and adjust their output accordingly. If there is a greatly increased demand for shoes, established firms increase their output and new firms enter the market; when demand falls, output is cut back and inefficient firms may fail. In the absence of monopoly, producers and purchasers of shoes reach a level of interaction that is profitable for the former and utilitarian for the latter. This is how classical economic theory views the ideal market. People are forced to act rationally by the discipline of the market, and society benefits.

For the oil industry, however, unregulated supply-and-demand economics brings chaos. This is true for two reasons. First, unlike the case in the shoe industry, the oil industry does not have control over its own resources. Shoes are manufactured; oil is discovered. No one knows how much oil there is in the earth or where it is located. It is possible to have a huge industrial agglomeration and yet find no oil, as with Britain before the North Sea discoveries of the 1960s. It is similarly possible to have only a primitive economy yet produce copious quantities of oil, as is the case with some Arab states. This means that, contrary to the tenets of classical economic theory, an unregulated oil industry is incapable of acting rationally. Despite all its efforts to adjust and plan, too few discoveries may starve it, and too many discoveries may drown it.[16]

Second, in a society recognizing the private ownership of property, the unusual physical qualities of oil introduce a dynamic that leads to both economic and physical irrationality. If the oil is produced too rapidly, the natural gas or water that forces it to flow may diminish, and much of the oil will be unreachable. Therefore, to produce the oil in an efficient and reasonable manner, a rate of extraction that does not waste reservoir pressure must be discovered by engineers.

In the case of a nation that produces oil with a public company, or in the case of an oil pool held by a single private owner, the producers will almost always extract the oil at a safe rate. When a field is thus unitized, the interests of the individual operator in maximum profit and the interests of society at large in maximum recovery coincide.[17] This occurs because looking at the reservoir as a whole almost inevitably leads to a rational approach to developing it. Even when there are multiple owners, if they are forced by a government authority to combine their interests and share revenues, the field will most likely be produced at close to its optimum rate.

In the United States, however, where many landowners and operators may have access to a pool, individual and social interests conflict. Oil will migrate for long distances, out from under one person's land to another's. If any one operator begins to produce, all the adjoining landowners must likewise produce, or their neighbor will suck up all the oil and gas from beneath their property. So, they must drill their own wells to keep from being drained. And, of course, the same logic that forces people to produce also impels them to produce as fast as possible, in order to protect themselves.[18] As more wells are drilled into the reservoir, and as the pace of production is stepped up, the ultimate recovery from the field decreases. The individuals' interests in protecting their underground wealth from drainage collide with society's interests in ensuring maximum recovery.

In the early years of the petroleum industry, this problem of competitive production was made worse by the American judiciary's method of dealing with it. When oil was first produced in Pennsylvania, there were naturally lawsuits involving drainage, in which landowners claimed that the oil beneath their land was being expropriated by others' wells. Pennsylvania judges, having no technical means of tracing the underground migration of the oil and seeing no precedent for such a question in the casebook, fell back on an English common-law doctrine known as the rule of capture.

Petroleum, they reasoned, is like wild game: it moves quickly from one person's property to another's. In legal terms, it is fugacious. The common law clearly stated that, though a deer might have been born and raised on one person's land, the instant that it jumped across a boundary to another's property it became fair game to the second owner. Thus with petroleum: though it might have rested beneath my land for a million years, as soon as it flowed to your property and out of your well it was yours. Although Texas judges never applied the rule of capture in quite such an unsophisticated manner, they retained the fundamental point that producers were not liable for the damage they caused others by draining from under their land.[19]

The rule of capture, thus extended, prevented landowners from recovering oil drained to another person's property. A compulsory unitization law would have solved this problem by forcing owners to combine operations and develop the field as a whole. But there has never been a compulsory unitization statute in Texas. Since they had no legal remedy from petroleum poachers, the only protection for landowners lay in drilling and producing as fast as possible.

Because of these economic, physical, and legal peculiarities, the petroleum industry, if left to itself, was forced both to commit competitive suicide and to squander its own basic resource. When many different landholdings lay over an oil pool, unregulated production soon turned into a pumping contest, with each individual drilling as many wells as possible and running them wide open.

This was the practice in Texas for a generation after Spindletop. In every reservoir, it resulted in a huge waste because it guaranteed a dissipation of reservoir energy. Once 5 or 10 percent of the original oil had been produced, the water or gas drive was expended and the wells stopped flowing. Texas is dotted with small towns like Borger that went from boomtown to ghost town in a few years, because the frantic pace of development in their fields used up the pressure in their reservoirs.

In addition, although the oil must be quickly produced under unregulated conditions, it cannot necessarily be sold as quickly, so it is stored. Any oil that sits in one place is likely to dissipate: it is subject to leakage, fire, and evaporation. To the immense underground waste of reserves caused by flush production, therefore, is added the smaller but still significant amount of aboveground waste.[20] Furthermore, in a large field this unbridled production brings up so much of the oil so fast that prices are driven down, thus possibly making the whole endeavor unprofitable for everyone.

Unregulated, competitive oil production is therefore irrational both for society and for most of the companies in the industry. The larger the field and the more numerous the producers, the more likely the potential physical waste and economic self-destruction. This irrationality is felt most acutely by the large integrated corporations. As huge enterprises involving very large investments in many parts of the world, they need a relatively stable, predictable supply and market. They must take a long-run view. A great flood of unexpected oil, threatening, as it does, to greatly undercut prices, presents integrated companies with the specter of disaster. If prices are too low, their entire structure of investment is endangered.[21]

In the history of the oil business, therefore, major companies have always tried to control the source of oil, so as to make their corporate actions predictable. Majors always support restricting the supply of oil to the market demand, that is, releasing only enough of it to prevent oversupply from undercutting prices. The public has often condemned this activity as evil monopoly, but it is a matter of life or death for the majors.

In contrast, in the 1930s, small independents and royalty owners, who made relatively minor investments and who operated on a more modest scale, normally had a short-run perspective. Independents might be able to make a fortune selling large quantities of fortuitously discovered oil at cut-rate prices, and small landowners often wanted their royalties as quickly as possible. Both groups had little interest in price stability.

As a result, independents and small landowners often tended to side with the general public prior to the Great Depression and to denounce the majors' prorationing plans as price-fixing. Before 1930, this conflict between large and small segments of the industry simmered but never boiled. With the Daisy Bradford well and similar contemporary discoveries in Oklahoma, Venezuela, and the Middle East, however, the conflict became too violent to contain.

DEALING WITH CHAOS

The East Texas field, as the leviathan of all Texas oil pools, caused the greatest irrationality in the history of the industry. To an impoverished farmer in the Depression, an oil well was a lifeline, and landowners eagerly signed leasing agreements with the thousands of legitimate and crooked promoters who swarmed into the Kilgore area. Drilling rigs were hauled in from everywhere possible, and wells were put down in pastures, near front porches, in flower beds, and over parking lots. Because the major companies controlled so little of the land, there was no general view of the number of wells needed to develop the field, and no restraints were placed on drilling or production. Less than a year after the completion of the Daisy Bradford well, the field was producing a million barrels of oil a day, and a new well was being completed every hour.[22] At the end of 1931, there were officially 3,540 wells in the field; at the end of 1932, there were 9,400; and, at the end of 1933, there were nearly

12,000.[23] In 1929, land in the Kilgore area had leased for perhaps $1.50 an acre; two years later, some parcels went for $3,000.00.[24] Kilgore became the greatest boomtown in the fabulous history of black gold.

But what was good for Kilgore in particular was bad for the oil business in general. Arriving just at the start of the Depression, with prices already shaky and demand slacking off and with production from some of the larger Oklahoma fields already a problem, the cascade of oil from East Texas threatened to ruin the industry. In October of 1930, high-grade oil was selling for $1.10 a barrel, but early the next year the price fell to $.25. The major companies tried desperately to shore up their position by declaring a price of $.67 for East Texas crude, but independent refineries ignored the posted price and bought oil at lower and lower figures.[25] By the end of 1931, East Texas oil was selling for $.10 a barrel, and the price subsequently dipped even lower.[26] Oil fields elsewhere in the country, especially in West Texas, were forced to shut down because they could not produce profitably in competition with East Texas.[27] Economic disaster threatened the major companies.

Thus arrived the idea of the necessity of government-administered restraints on production: prorationing. During the 1920s the notion of conservation of reservoir pressure through production controls had gained some currency among the majors, but little had been done to recommend it as a public policy.[28] In the later years of the decade a voluntary plan of prorationing had been instituted in a few West Texas fields, notably Yates, but in the absence of some means of legal enforcement this was barely satisfactory.[29] Now, with East Texas threatening to swamp the industry, the major oil companies, some of the larger independents, and conservationists united under one thought: for reasons of reservoir protection and industry survival, prorationing had to be instituted in Texas—and fast.

The form they hoped that production restraints would take is known as market-demand proration. This authorizes a government agency not only to suppress each field's output to a level that will protect the reservoir but to adjust the production from all the state's wells to a level that will just meet the current demand for oil, at the current price. There is no doubt that restraining production to the maximum efficient rate of output is a conservation measure. But there was in the 1930s, and still is, a controversy about whether market-demand prorationing is also defensible as a conservation measure or whether it is simply an economic expedient. This complicated question will be explored in chapter 6.

Because of the dangerous fecundity of the East Texas field, the first five years of the 1930s witnessed an involved and intense struggle to establish market-demand prorationing and decide upon the appropriate government agency to administer it. This proration war was waged on three different fronts: the technical, the legal, and the political. The results from one constantly affected the circumstances of the others, but the issues involved in each can be made clearer by discussing them separately.

Technical Confusion

At the beginning of the Depression, the science of petroleum engineering was very much less developed than it is today. Reservoir dynamics were only imperfectly understood, and such knowledge as existed among geologists, physicists, and engineers had not reached the general public or even most of the business side of the industry. Especially, most of the smaller independents could not or did not want to employ scientists; whatever curiosity they felt about oil pools was satisfied by drilling into them. They were consequently not prepared to respect the advice of a geologist employed by a major company that they were damaging the ultimate recovery of the field by running their individual holdings wide open.

Moreover, the counsels of scientists to respect the great field's fragility lacked force, because they themselves did not at first understand its propulsive mechanism. Prior to 1930, two types of drive were familiar to the technician of hydrocarbon production. In the first, natural gas was dissolved with the oil and pushed it out of the well pipe, much as the carbon dioxide pressurized in a cola bottle forces the soft drink out into the atmosphere when the bottle cap is removed. This is called dissolved gas drive. In the second, an excess of free gas sitting on top of the oil pushes it downward and thence up the well casing, much as a pump spray bottle forces hair spray up a tube and out. This is called gas-cap drive.[30]

The East Texas field, however, possessed a third, relatively unknown type of propulsion: water drive. Water, expanding all through the Woodbine sandstone for a hundred and fifty miles, pushed the oil into the eastern stratigraphic trap, originally at over sixteen hundred pounds per square inch of pressure.[31]

When the East Texas field was discovered in 1930, the mechanics of water drives were close to being a mystery. There was not even agreement among scientists as to whether such a thing actually existed. This technical ambiguity was all the more difficult to resolve because the search for scientific knowledge of the field was intimately connected to the political controversy surrounding prorationing. Also, to those who opposed regulation for reasons of self-interest, the confirmation of a water drive would be an economic threat because its existence would imply the need for production control.

Oil in the East Texas reservoir had about the same resistance to flowing, or viscosity, as kerosene. If the pressure on the east side of the reservoir were allowed to drop, the gas dissolved in the oil would come out of solution, and the oil would become more viscous. It would more closely resemble tar than kerosene. As the pressure continued to drop, water, pressured from the west, would flow through the sandstone toward the east. Because of the greater viscosity of the oil, the water would encounter more and more resistance as the "tarry" oil clung ever more stickily to the tiny pores in the rock.

The water would encroach more rapidly in the more permeable parts of the sandstone, more slowly in the less permeable sections. Eventually, the more rapidly moving water would cut off and surround the more slowly advancing oil, effectively stopping its flow and ensuring that it would never be brought to the surface.

If production were restrained, however, the gas would not come out of solution, the oil would not become more viscous, it would not stick more firmly to the stone, and the water on the west side would advance slowly and evenly, pushing the oil toward the east. A high percentage of the oil would eventually be recovered.[32] If a water drive could be proven to exist, therefore, production control to protect the field would be mandatory. As a consequence, those who opposed regulation were bound to refuse to accept the existence of a water drive in East Texas.

In the first years after the Daisy Bradford well came in, Commission personnel attempted to understand the forces at work in the field, but, because of the political implications of any technical conclusion, they encountered organized skepticism. To make matters worse, the Commission was not at first equipped to deal with the challenge of comprehending the field. In 1931, its employees did not know how many wells were in the field or who the operators were. They possessed no map of the field and had no data on its geological structure. Their technical uncertainty subverted the authority of the Commission, for if its agents could not understand the reservoir they could hardly claim credibly to be acting to protect it.

The Commission began to get a handle on the problem of information in late 1931. At the request of their East Texas engineering staff, Commissioners issued an order that every royalty owner had to furnish a record of the number of her or his wells and the operator of each well, plus a land survey. With this data, Commission engineers were able to construct a map of the field.

By December of 1932, the Commission was able to shut down the field while its engineers, using pressure gauges borrowed from the major oil companies, recorded bottom-hole pressure in twenty-eight wells, running from east to west. In the initial tests, pressure varied from fourteen hundred pounds per square inch on the west side to seven hundred on the east. Seventy-two hours later, when the same wells were retested, pressure on the west side was unchanged, but on the east it had almost doubled.[33]

This finding was significant because, in an ordinary gas-drive reservoir, pressure does not rise again once it has decreased. The only way to explain the rise in pressure on the east side of the field was by admitting that a water drive existed. The acceptance of a water drive implied that production would have to be regulated to protect the field. This scientific understanding of East Texas was thus the first great victory for conservationists in the proration war.

To those who opposed regulation for reasons of self-interest, of course, this engineering understanding of the field's dynamics was a threat. Some refused to believe the Commission's conclusions or endorsed competing scientific interpretations. As late as the spring of 1934, alternative explanations of the facts of the field appeared in local publications.[34] But, once the existence of a water drive was relatively well established, some kind of regulation was assured. The form that control would take, however, when it would be imposed, and the agency responsible for it were by no means certain. The technical struggle to understand the field thus became relatively less important than the legal and political conflicts that were occurring at the same time.

Legal Ambiguity

In the absence of a state statute to compel landholders to unitize their holdings, Texas courts applied the rule of capture to disputes dealing with drainage between tracts. A regulatory agency, therefore, could not force landowners to share their oil once it was out of the ground. Regulation would have to govern what was done with the oil while it was still buried. In trying to accomplish this, however, the Railroad Commission ran into a legal dilemma created by a combination of physical and economic circumstances.

Unrestrained production squandered natural resources. To try to conserve these resources was reasonable, legal, and even patriotic.

But unrestrained production also destabilized prices and encouraged competition. To try to control the market was monopolistic, illegal, and wicked. Advocates of production controls argued that prorationing would be a conservation measure and therefore legal, which was true. Opponents of control argued that prorationing would be a price-fixing scheme and therefore illegal, which was also true.

Proponents of production controls were not very forthright in their handling of this problem. Although it was clear to everyone that the main reasons for the sudden interest in prorationing in the 1930s were economic, its supporters usually insisted publicly that their motives were entirely disinterested.[35] The suspicion created by these claims poisoned both the legal and the political atmosphere. But that is beside the point. The fact is that prorationing is both a means of conservation and a stratagem for price-fixing.

In their efforts to convince the courts that prorationing was acceptable, Commissioners ran into another minefield. In the early 1930s, both state and federal courts often claimed the authority to closely supervise regulatory agencies. The Commission's job was made much more difficult by judges looking over its shoulder, so to speak, and invalidating its orders on both procedural and substantive grounds. Although later in the decade the courts retreated to a more tolerant position, during its period of maximum vulnerability the Commission found itself frequently second-guessed by judges, regarding not only whether it had the power to regulate or whether it had followed due process but even whether it had its facts right.

A third legal trauma was visited on the Commission before it got a grip on prorationing: the Texas legislature kept changing its authority. In 1919, the Commission was authorized to prevent waste.[36] In 1929, this act was amended to exclude "economic waste"; that is, the Commission was asked to control the physical effects of production but not the economic effects.[37] In August of 1931, the Commission was specifically forbidden to limit production to market demand.[38] Finally, in November of 1932, in the face of bedlam and insurrection in East Texas, the legislature reversed itself and authorized the Commission to prorate production to market demand.[39]

As a result of these pronouncements by the courts and vacillations by the legislature, the legal authority of the Commission was in doubt until over three years after the Daisy Bradford well was brought in. Every time the Commission attempted to control production, hosts of independent operators went to court to stop it, many obtaining injunctions and some winning important cases. Even when judges upheld the Commission, the legal uncertainty surrounding prorationing encouraged producers to evade or ignore Commission orders. "Hot oil," produced in defiance of the Commission's proration orders, flowed freely, creating or contributing to some of the most famous fortunes in Texas.

In April of 1931, the Railroad Commission issued its first proration order for East Texas, limiting each well in the field to about a thousand barrels a day.[40] Four months later, in the Macmillan case, a federal court struck down the order, on the grounds that it bore no relation to physical waste but was an effort to fix prices.[41] On August 12, the legislature passed an antiwaste law, but, reacting to the Macmillan case, it expressly forbade the Commission to prorate to market demand.[42] Five days later, with a million barrels a day flowing from the Woodbine, East Texas in a frenzy, and crude oil selling nationally for about 60 percent of its price a year earlier,[43] Governor Ross Sterling declared martial law and sent troops to the field.

In September of 1931, with Sterling's troops holding the field in a semblance of order, the Railroad Commission tried again, prorating East Texas to 400,000 barrels a day, allowing each well about 225 barrels without respect to its potential. In December, the Commissioners, seeing enforcement of the proration order under full control of the governor, withdrew from the fray, leaving the problem of administration to the military. But on February 18, 1932, a federal court held Sterling's action illegal and ordered an end to martial law.[44] So, on February 25, the Commission issued a proration order based on a flat per-well allowable of 75 barrels a day. In October, the federal court invalidated this order on the grounds that it violated the state anti-market-demand act of the previous year and that the per-well allowable necessarily discriminated against wells of greater potential production.[45]

The next month, with hot oil again spouting in East Texas, Governor Sterling called a special session of the legislature, which this time authorized the Commission to prorate to market demand.[46] On April 22 of 1933, the Commission issued a new order, which set the East Texas allowable rather high, at 750,000 barrels a day, but distributed allowables among wells partly on the basis of their bottom-hole pressure. In January of 1934, at long last, a federal court upheld this order, and the legal authority of the Commission was established.[47]

But, by this time, legal authority was not enough. During the struggle for legal clarity, hot oil running had become a way of life in East Texas, and there had been efforts in both Austin and Washington to solve the problem of production control by taking it away from the Commission. This struggle touched on the judiciary, but its principal action was on the stage of politics.

Political Dilemma

By the beginning of the 1930s, the post of Railroad Commissioner had become largely a sinecure for over-the-hill Texas politicians. The work of regulating intrastate transportation was neither physically nor mentally taxing, but it provided an opportunity for men who were past their prime to dispense patronage and generally remain on the fringes of state government. It was customary for a Texas governor to appoint an older man who had served the Democratic party faithfully to a position vacated by another older man. The new appointee would then have the advantage of incumbency in the next election. When the Daisy Bradford well came in in 1930, the three Commissioners were scarcely ready to confront a problem demanding energy and imagination. These three—Charles V. Terrell, sixty-nine years old, a former judge, state treasurer, and state legislator; Lon Smith, sixty-one years old, a former state legislator and comptroller; and Pat Neff, fifty-eight years old, a former governor—responded to the pandemonium in East Texas by ignoring it for seven months.[48] By the time they issued their first proration order in April of 1931, the situation was far out of control.

Their lack of vigor was complemented by the agency's traditional personnel policies. The Railroad Commission was the very embodiment of the spoils system. Each Commissioner controlled one-third of the employees. Whenever there was a change of Commissioners, a third of the agency's employees would be fired and a new group hired. These employees were in no sense technically qualified to help regulate the petroleum industry. They were either friends of a specific Commissioner or people who in some manner had been able to pull strings in their local communities. In the Texas State Archives, there are boxes of letters written to Commissioners during this period, asking for employment or endorsing someone else's request for a job.[49] The usual procedure called for some lad to approach a prominent citizen—often a judge or a business owner—of the nearest small town and ask for a letter of recommendation. These letters invariably praised the boy's winning personality and loyal character; they rarely mentioned whether or not he knew anything about oil or railroads.

Such a personnel policy had the inevitable result: incompetence. The Commission employed no petroleum geologists and only one engineer in 1930, and consequently its agents could not at first intelligently evaluate any of the technical arguments surrounding regulation or recognize violations of a Commission order. This hardly contributed to the authority of the Commission. Furthermore, stupendous fortunes were being made in East Texas, and the salary of a field inspector in the early days was less than two hundred dollars a month. Thus the job of Commission inspector became, as George Sessions Perry remarked, "fraught with the most paralyzing temptations."[50] It did not help the agency's reputation when the attorney son of one Commissioner was allowed to practice before the Commission,[51] or when the son was subsequently indicted for attempting to bribe a Commission employee,[52] or when another Commissioner allowed his chief enforcement officer in East Texas to take a leave of absence to engage in some profitable petroleum-related business ventures.[53]

This is not to say that all Commission employees were always technically unqualified or that many of them were dishonest. As the decade advanced, Commissioners hired inspectors with excellent technical training. But the Commissioners refused to give up the patronage system, and that reluctance cost them dearly. In 1934, chief East Texas inspector R. D. Parker was released from his job for firing incompetent and/or dishonest employees and replaced by a man who would not rock the patronage boat. Such infusions of politics into the attempted administration of prorationing undermined respect for the Commission on both state and national levels.[54]

That respect was urgently needed, for the Commission was attempting to control a virtual war that began to rage in East Texas in 1931 between the advocates and the opponents of production control. Major integrated companies, large independents, large landowners, and conservationists supported market-demand proration. Independent refiners and the very numerous smaller independent producers and royalty owners opposed it. The struggle was always intense, frequently acrimonious, and sometimes violent. Supporters of prorationing had better financing, scientific arguments, and national political figures on their side. Their opponents had many more votes, the general public's suspicion of large corporations, and the difficulty of enforcement on theirs.

The predicament facing any state regulatory agency was this: how to enforce a measure that was vitally important in the long run on a large and resourceful population whose short-run interests directly contradicted that measure? As political animals, Commissioners were faced with the delicate task of attempting to suppress an activity that thousands of Texans viewed as their economic salvation without committing political suicide. The first great policy decision of the Commission thus involved a seemingly insoluble dilemma.

As with any political battle in the United States, the proration war was composed of a series of battles fought through organizations and publicity. The majors, who early recognized that they were quite generally hated, sought to work through sympathetic independents. As W. S. Farish, chairman of Humble, later recalled, "We have learned from experience . . . that one of the easiest ways to defeat a thing is for us to ask for it."[55] In 1932, the Texas Oil and Gas Conservation Association was organized to promote prorationing. Its membership was almost entirely composed of respected independents, but its technical advice and finances came largely from the majors.[56] Majors and those independents against hot oil published magazines in East Texas designed to spread the word about prorationing: the *Conservationist, Independent Petroleum World, East Texas Oil.*

Opponents of prorationing were just as busy. They had their own magazine, the *Independent East Texas Oil and Gas News,* and, at first, the editorial support of most local papers. In pamphlets, articles, and speeches, they hammered away at a single theme: prorationing is monopoly and price-fixing foisted on the industry by selfish majors who have no concern for the little guy.[57] They too organized trade associations to protect their interests, the most important of which was the East Texas Lease, Royalty Owners and Producers Association. And they ran hot oil. By some estimates better than one hundred thousand barrels a day of unauthorized petroleum left East Texas during portions of 1932 and 1933.[58]

Independents at first tended to view the efforts of any government agency to control the field as part of a conspiracy inspired by the majors to eradicate small producers. Their suspicions were not allayed by the backgrounds of some of the prominent actors in the proration drama. Governor Sterling, who declared martial law in August of 1931, had formerly been chief executive of Humble. The man he put in charge of the military enforcement was General Jacob Wolters, who in civilian life was chief Austin lobbyist for the Texas Company (now Texaco).[59] One of Wolters' colonels was a production executive for Gulf.[60] Some Commission engineers had formerly been employed by a major company. In June of 1932, when Pat Neff resigned, Sterling appointed Colonel Ernest O. Thompson to the Commission. Thompson had been vice-president of an organization that supported prorationing and as such was suspected by many in East Texas of being a front for the majors.[61]

They were wrong about Thompson. He was not a spokesman for any particular viewpoint or set of interests. Thompson was a great politician—he was finely attuned to the social forces that would somehow have to be reconciled if peace were to come to the Texas petroleum industry. To the colonel would belong much of the credit for solving the political dilemma facing the Commission and thus preventing the Texas petroleum industry from destroying itself.

Thompson was born in Alvord, a small town in Wise County, North Central Texas, in 1892. He attended Virginia Military Institute and received a law degree from the University of Texas in 1917.[62] He had been a highly decorated commander of a machine-gun battalion in World War One and a remarkably successful mayor of Amarillo.[63] In all the episodes of his life, he had displayed ability, intelligence, self-confidence bordering on arrogance, and an almost charismatic touch in interpersonal relations.

That the colonel understood the nature of the conflict over prorationing is revealed by the hundreds of speeches, press releases, and pamphlets he produced during the almost thirty-three years that he served on the Commission. Thompson grasped the problem facing the Commission: to stabilize the industry and promote conservation, the insitution of prorationing was a necessity, but it must be done in a manner that reassured the independents and royalty owners that they would not be exploited by the major companies. Prorationing might be imposed by force, but, until it acquired the assent of the small producers, the Texas oil industry would remain in civil war.

Thompson thus set out to establish prorationing under conditions that would make it clear that the independents would be protected. The first goal he began to achieve by taking personal charge of the Railroad Commission field office in East Texas; the second would be accomplished by launching a private campaign to recruit independent support. This latter effort was carried on both positively, by praising the independents, and negatively, by attacking the majors. The tone of this political strategy is well represented by a statement he released in announcing his candidacy for a full term in 1935:

> The issue is not one of big business vs. the small man. It is greater than that—the real issue is shall Texas youth continue to have free opportunity amid their natural resources? Or—

> shall our youth in the near future be forced through lack of opportunity to work for some big company in a chain filling station?[64]

It would be inaccurate to argue that Thompson united the industry by providing the majors with something tangible—prorationing—and the independents with only symbolic reassurance. As the following chapter will demonstrate, the Railroad Commission took quite concrete steps to ensure the viability of the small producer and landowner. But the prerequisites to an acceptance of prorationing were comity and mutual tolerance in the Texas industry. Fashioning these was the function of political leadership and would not have been possible without Thompson's ability to create trust in the Commission.

Before Thompson had mastered the situation, however, the Commission had to survive three close calls, during which it very nearly lost its authority. In March of 1933, with hot oil still a major problem in Texas, the new Roosevelt administration was inaugurated in Washington. Nowadays, members of the petroleum industry are known for their hostility to meddling from the national government, but in the first spring of the New Deal a majority of oil and gas producers were willing to endorse the establishment of a federal "oil czar" to impose order.[65] The idea was acceptable to both President Roosevelt and Secretary of the Interior Harold Ickes, who would have been the man responsible for controlling production. It would have been passed easily by a compliant Congress. But independents and majors could no more agree on the details of federal control than they could come to terms on the problem of East Texas.

While a sympathetic Sam Rayburn tied up the federal control bill in his House Committee on Interstate and Foreign Commerce, a small group of independents convinced Ickes and Roosevelt that integrated companies were dangerous and that, therefore, any federal legislation should force the majors to divorce themselves of pipelines. FDR sent a letter endorsing this position to the governors of seventeen petroleum-producing states on April 3.[66] Since divorcement meant the dismemberment of their companies, the majors immediately reversed their position on the desirability of federal control. With Rayburn's committee hostile to the bill and with the majors opposed, the immediate possibility of an oil czar died a quick death.[67]

Instead of imposing control from Washington, the New Deal next supported state prorationing. In July of 1933, under authority granted him by the National Industrial Recovery Act, Roosevelt issued an executive order banning the transportation in interstate commerce of petroleum produced in violation of state statutes.[68] Federal agents were sent to East Texas that fall, and, after a long and administratively complex series of maneuvers, they succeeded in aiding the Railroad Commission in shutting down the hot oilers. Prorationing was temporarily effective.[69]

Meanwhile, however, the Commission had nearly lost its authority on the state front. In the spring of 1933, critics of the Commission argued that it was inefficient because its members were elected; too much democracy, they contended, made for a timid, dependent agency. Although Thompson and his allies lobbied strenuously in Austin to counter this assumption, opinion in the capital was running strongly in favor of fashioning another agency to handle the oil crisis. Governor Ferguson, all the majors, and a majority of the independents supported a bill that would have stripped the Railroad Commission of its petroleum-regulating powers and transferred them to a newly created, appointive Natural Resources Commission.

The bill narrowly passed the Texas House on May 2 and was expected to breeze through the Senate the following day. But that evening Representative Gordon Burns, a leader in the opposition to the bill, was accosted and badly beaten in the lobby of a downtown Austin hotel by four anti-Commission independents. The news spread quickly and outraged public opinion. To support the new Natural Resources Commission now, it seemed to senators, would be to brand themselves as frightened and driven by the oil lobby. The next day, with Burns watching from the floor in a wheelchair, the Senate voted down the bill twenty to ten, thus saving the Railroad Commission.[70]

In the second half of 1934, there was again the possibility of federal control over the oil industry. The Thomas-Disney bill, introduced in Congress in the spring of that year, was essentially the oil czar measure of the previous session. The opposing coalitions in this struggle were more scrambled than in the previous year, partly because the majors by this time had learned to work through independents and partly because the opposition to the bill was better organized and more outspoken.

Despite the fact that representatives of independent producers supported the bill and that there is evidence of some opposition by the majors,[71] the principal tactic of those opposed was again to cast the bill as an effort by the integrated companies to gain control of the industry. Typical of independent arguments was a telegram sent by L. H. Gray, president of a small Texas refinery, to Senator Tom Connally in November of 1934:

> . . . suggest that you use grave consideration before supporting the Thomas or other bill jeopardizing Independents interests stop don't consider too seriously telegrams from Texas Chambers of Commerce which are controlled by major companies . . .[72]

As with the state struggle, Ernest Thompson was at the center of the fight to forestall federal intervention. He was in constant contact with pro-Commission oilmen in Washington via telegrams and letters, and he made dozens of speeches in Texas to industry gatherings. Over and over, he repeated his two themes: prorationing was a conservation measure that was in the long-run interests of majors and independents, large and small landowners alike, and only a state agency was close enough to the oil problem to deal with it competently.[73]

The Thomas-Disney bill probably would have died under Sam Rayburn's unfriendly hand in any case, but in November of 1934 Secretary Ickes effectively destroyed its chances during a speech to the American Petroleum Institute in Dallas. Accusing Texas in particular of being lackadaisical in its prevention of waste, Ickes implied that the time had come to think of the oil industry as a public utility and to regulate it accordingly.[74] The response was immediate. The API, which had sent representatives to testify for the bill, promptly withdrew its support;[75] the independents launched further diatribes;[76] and the bill perished. The Commission had once again survived a threat.

By the beginning of 1935, the combination of Commission and federal policing had become effective, and the hot oil problem was under control. In January, however, the United States Supreme Court declared section 9c—the hot oil provision—of the National Industrial Recovery Act unconstitutional, and state prorationing again teetered on the brink.

In a masterpiece of legislative efficiency, however, the Connally Hot Oil Act was passed by Congress little more than a month later. This law prohibited the movement in interstate commerce of illegally produced oil, as had section 9c, but was written to meet the constitutional objections of the Supreme Court. Further, it authorized the states to establish an organization to coordinate their prorationing plans. All important oil-producing states but California soon complied, and the Interstate Oil Compact Commission was born. The Connally Act passed court tests and became the framework within which the Railroad Commission and other state agencies regulated domestic oil production until 1972.

The defeat of the Thomas-Disney bill, the passage of the Connally Act, and the successful enforcement of prorationing were all possible because by 1934 the Railroad Commission had convinced most independents that regulation was in their interests. The swing in independent sentiment is nicely illustrated by the career of Carl Estes. During the early days of the East Texas boom, Estes, as editor of the *Tyler Courier-Times,* had become famous for intemperate editorials denouncing major companies, praising small operators, and opposing prorationing. He had been secretary-treasurer of the most important producers' association formed to fight prorationing.[77] By early 1934, however, Estes had become editor of *East Texas Oil* and was writing equally intemperate editorials endorsing prorationing and damning hot oil runners.[78] With such a swing in independent opinion, the political battle of the Railroad Commission was won.

This acquiescence by the independents extended only as far as producers and royalty owners. Small independent refiners, who depended absolutely on large quantities of cheap oil, never accepted prorationing—and rightly, for most of them perished when Commission regulations became effective.[79] The overwhelming majority of independents, however, made their living in production, as opposed to refining, so the extinction of the small refineries did not threaten their self-interest.

In the campaign that created public confidence in the Commission, Ernest Thompson was unquestionably the central figure. In a multitude of speeches, he conveyed the message that the Commission was willing to fashion a compromise: it would restrict production, as the majors wanted, but in other respects it would favor the independents. It was not possible to make such an arrangement believable with words alone. As the following chapter will detail, the Commission ensured acceptance by adopting a policy of spacing and allocation that clearly favored the independents. The majors might complain, but they had won the most important battle.

CONCLUSION: THE BEGINNING OF REGULATION

In the first great policy-making episode of the Railroad Commission, much of the action took place outside the agency, in courts or legislatures. That the Commission retained authority over Texas oil production was largely due to fortune. The Commission was nevertheless the only body able to comprehend and master the political divisions that were rending the industry. The credit for fashioning the compromise that brought peace to the Texas oil and gas business goes largely to Ernest Thompson. Despite the fact that he was an appointee, Thompson was electorally accountable, and he succeeded in uniting the industry while knowing full well that he would have to answer to the voters. Thompson handled the two great contending forces of the proration battle by giving each side the half a loaf that was most vital to its survival. In doing so, he and his fellow Commissioners solved the overproduction problem but created a host of others. They were immediately forced to deal with the consequences of the prorationing policy on other fronts. Their successors deal with some of those problems yet.

3. The Ironies of Regulation, 1935–1950

Nothing can be more dangerous than the impact of private interests on public affairs.—Jean Jacques Rousseau, *The Social Contract*

Winning the authority to prorate oil production marked the beginning of the Railroad Commission's ascent to a position of paramount importance in United States petroleum policy making. From 1935 to the 1970s, national oil policy was created largely in the states, and the Railroad Commission was by far the most important of the state agencies. With Texas possessing just under half of the nation's reserves, Commissioners exercised decisive control over the supply of oil.

But oil prorationing was only the most obvious of Commission responsibilities. Once that basic power was won, a host of subsidiary issues arose to occupy the attention of Commissioners. In the days of flush production, the limits of each well and field were set by nature; when humans chose to replace nature, someone had to decide how much each operator would be allowed to produce. By accepting the *general* responsibility for the regulation of oil production, therefore, the Commissioners had the *specific* authority for overseeing every well in every field thrust upon them at the same time.

For example, any government agency that prorates production must decide how to allocate output among fields. Once a state maximum total is decided upon, Commissioners must choose among a variety of alternative methods of dividing the pie. Should the large, rich fields be permitted to dominate the state total, or should their output be cut back somewhat so that the smaller fields may be more economically viable? Should fields along the coast or otherwise close to transportation facilities be allowed to produce at high rates, or should their allowables be set so low that purchasers would be forced to build pipelines to all the fields of the state? Should fields owned by certain people or interests be favored, or should rule makers ignore the economic and political consequences of their decisions? Should allocation formulas differentiate among crude oils of varying chemical composition, or should they treat all oil the same?

Attempting to deal with questions such as these, the Commissioners gradually evolved their allocation policy during the 1930s and 1940s. At first, only the East Texas field was prorated. In the late thirties, the Commission began to regulate all the fields in an area together. During World War Two, it combined this area regulation with proration by the type of crude oil needed for the war effort. Finally, in 1950, the Commission adopted the yardstick method of prorationing and applied it across the board to most fields in the state.

The yardstick adopted in 1950 had originally been created in 1947 for the purpose of assigning temporary allowables to new wells. It consisted of a set of rules decreeing maximum production limits for fields by their depth and well density. If a field were located at two thousand feet, for example, and its wells were spaced twenty acres apart, each of its wells was given a theoretical maximum of 55 barrels a day. If the field were at five thousand feet on forty-acre spacing, each well had a theoretical maximum of 102 barrels a day, and so on.[1] Yardstick schedules were published and enabled producers to calculate the maximum output permitted for each of their wells.

Under prorationing, of course, the wells in a field were almost never permitted to produce at their theoretical maximum, whether that maximum was assigned by studying the maximum efficient rate of production or by applying a yardstick. In every year from the end of World War Two to 1972, except 1948, the Commission cut back actual production to some percentage of the theoretical maximum, in order to match the state's output to the market demand.

Although there has not been much economic investigation of the consequences of the 1947 yardstick,[2] the general assumption in the industry is that it benefited low-production fields at the expense of flush fields.[3] As a rule of thumb, independents tend to be concentrated in poorer-quality fields. The yardstick system of establishing proration maximums thus may be one of the political payments that the Railroad Commission made to the independents for ceasing to obstruct prorationing.

The evolution of statewide allocation involved political disagreements, but, compared to the original proration battle, it was relatively peaceful. Other problems, however, caused more trouble. Attempting to ration production, guarantee access to transportation facilities, and foster the state economy at the same time, Commissioners faced a sequence of difficult policy choices, each of which aroused intense political pressures from all or parts of the industry. As these struggles unfolded, they revealed a series of great ironies, ironies that illustrate some of the paradoxes of regulation in America.

One of these involved the shift in support for market-demand prorationing. In the early 1930s, the major integrated companies had been the advocates of prorationing, opposed by the smaller independents and landowners. As the implications of government control of production became clear, however, the axes of support for prorationing revolved. The major companies wished to prevent the kind of gush of oil that had undercut prices in the 1930s, but they did not otherwise encourage Railroad Commission interference in their autonomy. Statewide market-demand prorationing, however, by limiting access to markets, forced the majors to accommodate themselves to independent needs.

On those few occasions when the Commission set its statewide allowables above a level desired by major purchasers, the majors instituted pipeline prorationing; that is, they took the extra oil from their own wells in their own fields and ignored the independents.[4] They did this because it was cheaper and more convenient for them; the fact that many independents might go bankrupt for want of an outlet for their oil was of little importance to the majors. But by estimating statewide market demand precisely, as it usually did, the Commission ensured that the majors would have to connect their pipelines to every independent producer. Market-demand prorationing, originally instituted to protect the majors from price-cutting by the independents, became a means to force majors to provide independents with a market. Additionally, as will be shown shortly, in setting specific rules of production the Commission invariably favored the interests of the smaller segments of the industry over those of the larger.

By the late 1940s, therefore, the major companies had grown less supportive of the Commission, but the independents were embracing it enthusiastically. A system that had initially been forced on the independents almost at the point of violence turned out to be the best thing that ever happened to them.

Another great irony involved the clash of individual and group interests. Although by the late thirties people in the petroleum industry had almost unanimously accepted state control of production in principle, most of them could be counted on to oppose a specific application of regulation whenever it clashed with their own interests. Commissioners discovered that they were forced to impose long-run policies on a mass of individuals who were governed almost entirely by short-run perspectives. The Commissioners developed increasingly convoluted rules in an attempt to manage the opposition of various interests to specific parts of their general policies.

Of the policy decisions that were thrust upon the Commission in the first two decades of its responsibility for conservation, those that inspired the most conflict and had the greatest ultimate consequences pertained to the establishment of rules for spacing and allocation within fields and to the campaign to outlaw the flaring of natural gas. In the first of these conflicts, the Commission kept the tacit bargain it had made with the small producers and landowners in the proration battle and sacrificed economic efficiency to their interests. In the second, the Commission rose above political pressure and, in a series of statesmanlike decrees, won a great victory for conservation.

POLICY-MAKING EPISODE NUMBER TWO
Spacing and Allocation, Round One

Ever since 1919, the Railroad Commission had been responsible for regulating the spacing of wells. But, in the 1920s, these rules had been so relaxed, and their enforcement had been so loose, that they would be more properly described as a potential rather than an active power of the Commission. By assuming the task of controlling the state's oil output in the 1930s, however, the Commission made the spacing rule a vital cog in the regulatory machinery. Rules governing the spacing of wells were intimately connected to formulas allocating their production—the more wells on any given area of land, the more oil could be produced on that land. As greater production meant greater wealth, the Commission's spacing and allocation policies were the object of the greatest possible industry interest. The political battles over spacing and allocation policy were violent and confused, partly because billions of dollars depended on Commission decisions and partly because the standards to be applied to such policy were inherently unclear.

POOLS, PRODUCERS, AND CONSERVATION

The allocation and spacing problem stems from a basic difficulty over the meaning of the concept of conservation. The Railroad Commission, by statute and political predisposition, was charged with preventing both physical and economic waste. It was thus a conservation agency, as its publicists never failed to emphasize. But the definition of conservation is ambiguous and can be interpreted in different ways, depending on the interests and perspectives of the interpreters. These different interpretations lead to divergent policy recommendations on matters of well drilling.

There is no disagreement about the actual physical loss of oil that was a component of the call for prorationing. Any measure that stops the rapid destruction of reservoirs and keeps already produced oil from running down creeks, evaporating, and, in general, being wasted is a conservation rule. But that is only part of conservation. After the prevention of physical destruction, the components of conservation become obscure and arguable.

Economists have a definite perspective on the notion of conservation. Although they differ somewhat among themselves in wording and emphasis, all argue that conservation involves maximizing the present value of a resource. In practical terms, this means getting the most petroleum for the least investment. The general formula is subject to modifications for rate of production; that is, it may be permissible, in a conservation sense, to drill more wells and get the oil faster rather than fewer wells and get it slower, because that generates a more rapid return on investment. Generally, however, the attitude of economists is well represented by an understanding that defines conservation as the prevention of "avoidable expense in the developing and producing of crude oil."[5]

The adoption of this definition of conservation has definite implications. If oil is to be produced for the smallest investment, the minimum number of wells that can drain any given field should be drilled. In order for this minimum of drilling to be implemented, the reservoir must be treated as a whole. This is so because, where there are multiple owners over a pool, they must cooperate for economically efficient development. If all owners but one drill few wells, and if that individual drills many, that person will drain oil from under their land. To protect themselves, they will in turn drill more wells, and conservation will not be achieved. So, in the interests of all, the economic perspective recommends unitization—developing a reservoir as a single entity and sharing the revenue among producers—for maximum conservation.

Economists always recommend unitization as the best method of producing the most oil for the least investment and, therefore, as the surest road to conservation.[6] As a corollary, they suggest the widest efficient spacing between wells. Because major oil companies normally have large investments on extensive parcels of land, they almost invariably share the economists' perspective on conservation. As a result, they have, since the thirties, usually been eager to unitize operations in every field in which they participate. Where unitization is not practical, they urge regulatory agencies to enforce wide-spacing rules.[7]

Engineers also normally recommend unitization but for somewhat different reasons. When reservoirs are developed as entities, technical means can be applied that will lead to a larger recovery of oil than would result from fragmented development. For example, under a secondary recovery project, the wells at the periphery of a field can be used to inject gas or water into the rock formation, thus greatly increasing the output at the center of the field. Clearly, unless the operators of the wells on the edges of the field agree to give up their own production and the operators near the center agree to share their revenue, such a project is impossible. Secondary recovery can often double the ultimate output of a field, and so cooperation is manifestly in the interests of conservation.[8] Engineers invariably support unitization because, by permitting secondary recovery, it increases the amount of retrievable oil.

Small operators and landowners, however, often have a different perspective on the meaning of conservation and, hence, on unitization and spacing. For both ideological and financial reasons, they often prefer to develop their property individually; this leads them to oppose unitization. Many simply don't trust someone else to handle their business; many are hostile to the major companies; some feel that they can make a greater profit by holding out until others in a secondary recovery project increase their share of the return.[9]

Because small operators and landowners prefer individual development, they also tend to support narrow spacing. A multitude of small tracts over a field naturally makes for many closely drilled wells, and the smaller the tracts, the closer the drilling. In the absence of unitization, small operators argue that this "drilling it like a sieve" is in the interests of conservation, because such development ensures a maximum recovery of the oil in place. They maintain that an emphasis on economic efficiency leads to conclusions that make for so few wells that some oil may not be recovered. Conservation, these producers and landowners argue, should be thought of as the retrieval of the maximum amount of petroleum consistent with private property rights. If more oil will be recovered with more wells, then more wells should be drilled.[10]

As a result of these differing perspectives, the assumption of prorationing automatically created a conflict for a state regulatory body like the Railroad Commission. While everyone in the industry expressed *general* support for production control, many small producers *specifically* agitated for more, and more closely spaced, wells, all the while violently opposing unitization. The number of wells, of course, affected the allocation given to each well, so that the more wells, the less each one could be permitted to produce. As it prorated production in each field, therefore, the Commission became enmeshed in a struggle between small and large producers and royalty owners over allocation and spacing rules.

STYLE AND SUBSTANCE

The evolution of the Commission's policies toward spacing and allocation was affected by the personalistic, informal style of decision making that prevailed in the first decade of prorationing. Under the statute of 1919 that first made it a conservation agency, the Commission had developed a general spacing formula known as Rule 37. This rule, which continued, with some amendment, during the twenties, restricted oil wells to a density of one per two acres, with each well having to be positioned at least 150 feet from a property line and 300 feet from another well.[11] Exceptions to this rule could be, and frequently were, granted by single Commissioners, however. This rule was in force when the Daisy Bradford well came in.

In the early days of the East Texas boom, many operators simply ignored the spacing rule and drilled where they pleased; those who bothered to go to Austin to apply for an exception were never turned down. In 1933, the Commission adopted a procedural rule declaring that every application for an exception to Rule 37 required a hearing. The intent of this change was not realized for years, however, because of the pervasiveness of an informal style of policy making on the Commission and because of personal conflict among the Commissioners.[12] Throughout the 1930s and into the next decade, the Commissioners attempted to create a rational, fair policy with regard to spacing and allocation, but they were inhibited by their own informal traditions and by their political subservience to the small producers.

For much of the decade of the 1930s, personal influence dominated the granting of Rule 37 exceptions. Prior to 1934, an operator desiring to drill would approach, or have a lawyer approach, any two of the Commissioners and informally argue the case for a permit. The appeal was likely to be well received, resulting in a speedy and satisfactory hearing. In 1934, however, Ernest Thompson and Lon Smith had a falling-out, and their intense enmity in the subsequent years colored all Railroad Commission activities, including the granting of "Rule 37s."

After 1934, applicants wishing to drill had to convince Thompson that their case was a good one. If Thompson approved an exception, Charles Terrell would automatically approve it and Smith would just as automatically refuse to support it. As the Commission was a body governed by majority vote, the assent of Thompson and Terrell was enough to carry the day. Woe to any ignorant operators, however, who secured the approval of Lon Smith first, for they had no hope of recruiting the other two and therefore no chance for a permit.

Not all the personal requests were granted by Thompson, but he displayed a marked tendency to be sympathetic to a small operator's request for a well. This was the result of the political compromise over prorationing: majors got production control, while independents got far more wells than were necessary to develop a field and allowables to make them profitable. The major companies remonstrated in vain with the Commissioners for granting so many permits. The little guy was the backbone of the constituency of the Commission, and the economic interpretation of conservation was secondary.

Thompson's domination of the Commission lasted until 1939, when Jerry Sadler replaced Charles Terrell. Sadler, who quickly came to dislike Thompson, teamed up with Lon Smith. Now the tables were turned, and a vigilant operator or lawyer had to wangle a permit by avoiding Thompson and cultivating the other two.

After Olin Culberson succeeded Smith in 1941, however, this highly personalistic system was somewhat modified, for, although Culberson did not get along with Thompson, neither was he particularly fond of Sadler. Without a cohesive bloc of two on the Commission, the method of granting Rule 37 permits became more formalized by default. This more impersonal approach persisted when Beauford Jester replaced Sadler in 1942.

The modification of procedure that occurred with the arrival of Culberson did not extend to the policy-making direction of the Commission. Culberson was, if anything, even more sympathetic to the small producer than was Sadler or Thompson. By all accounts, he was temperamentally at one with the small independents and normally took their side in any dispute within the industry.

But Culberson's concerns, like Thompson's, were more far-reaching than a desire to protect small operators. Like Thompson, Culberson wanted to spread the wealth inside Texas. Efficient production, leading to dominance by the large producers, especially the majors, would not necessarily create prosperity at home. Such efficiency, leading to a greater share of the industry for the majors, would result in fewer jobs in Texas and fewer royalty payments to Texas landowners, because the number of wells and the number of leases would contract. The major companies, even those that had been born at Spindletop, were controlled by out-of-state capital. As the most important goal was to keep oil money in Texas, it was legitimate to support inefficient production. Culberson articulated the concerns that dominated his thinking in a radio address a year after taking office in 1941:

> The combined petroleum industry of Texas pays annual wages and salaries of $272,000,000 to Texas workers. It pays $128,000,000 in lease and royalty money to landowners, $160,000,000 in purchase of equipment and supplies, $95,000,000 in local, state and federal taxes, and $95,000,000 in other expenditures, making a total of $750,000,000. . . . At least a million of our citizens or approximately one-sixth of the total population of Texas depend directly on employment in oil production, refining, or distribution.[13]

If such a state economic level were to be maintained, an allocation and spacing policy had to be pursued that created the maximum opportunity for drilling and production by small, homegrown operators. Culberson's arrival on the Commission consequently altered the style but not the substance of its approach to Rule 37 exceptions.

THE PRICE

The economic consequences of the relatively free and easy granting of Rule 37 exceptions in the 1930s were immense. As with so much else, East Texas set the tone for the rest of the state. By 1935, twenty-nine major companies owned 10,410 wells in the reservoir, and more than a thousand small firms owned over 12,000 wells. Approximately 65 percent of these latter wells were Rule 37 exceptions.[14] Officials for Humble estimated that the company had been forced to spend $5,500,000 in 1934 alone on economically unnecessary wells, to protect its leases from drainage by neighboring small tracts.[15] In 1937, two committees of the American Petroleum Institute reported that in their estimation unnecessary drilling in East Texas had cost the industry about $200,000,000.[16]

East Texas clearly illustrated the relationship between spacing and allocation. Although the general spacing rule for the field had been one well per ten acres since the early 1930s, so many Rule 37 exceptions had been granted that by 1939 there were over twenty-five thousand wells in the field, resulting in an average density of about one well per four acres. There were many examples of from five to ten wells on one-acre tracts, and one particular single-acre tract in Kilgore contained twenty-seven wells. The field's allowable was just under five hundred thousand barrels a day, which when distributed gave each well a maximum daily production of less than twenty barrels.[17] Thus, in the most prolific field in the world, wells capable of producing ten thousand barrels a day were restricted to one-fifth of one percent of their open-flow potential, while other, equally futile wells were constantly being drilled around them.

EVOLVING POLICY

Against this background of economic waste, and within the confines of their own determination to favor the small producer, Railroad Commissioners struggled to devise a workable policy for within-field spacing and allocation. The general doctrine endorsed by the courts, the legislature, and the Commission itself held that anyone owning a parcel of land or a petroleum lease, no matter how small, was entitled to one well to forestall "confiscation" of the property.[18] But this created an incentive to divide leases into tiny pieces, in order to get a well on each one. So, early in the 1930s, the Commission began to use the "illegal subdivision" standard for granting Rule 37 exceptions.

No lease that was both smaller than an owner's tract and smaller than the normal spacing rule permitted would be given a permit. Landowners possessing small tracts would be granted a permit to drill only if they had bought the land before the discovery of oil. If, however, the subdivision had occurred after the oil strike, only the original parcel of land would be considered in granting permits. This formula was upheld by the courts in November of 1937.[19]

The subdivision rule, of course, did nothing about the hundreds of small tracts that had existed in East Texas when the field was discovered. The wells on these drove the large landowners frantic trying to protect their oil. If, for example, a farmer owned a forty-acre bloc in East Texas, it was entitled, by the ten-acre density rule, to four wells. But if that bloc were surrounded by, say, twenty half-acre leases, each with a well, the large landowner was sure to be drained by these twenty wells. Only many more wells, drilled on the edges of the larger property, would stop the oil from migrating beyond its boundaries.

To be fair to the larger owners and producers faced with this situation, the Commission adopted the "equidistant offset" rule. Any producer could have a permit to drill an extra well as close to that producer's property line as a neighboring tract's well. A large lease touching several small tracts was therefore likely to be permitted more than its "correct" density of drilling, according to the spacing rule.

But the application of this rule meant that large leaseholders and landowners who did not abut small tracts, and hence were not entitled to equidistant offsets, were put at a disadvantage to others who did border small tracts. The leaseholder with the hypothetical forty-acre bloc surrounded by small tracts might end up with as many as twenty-four wells; a producer on a tract of identical size that did not touch many small tracts could drill no offsets and hence was allowed only four wells under the ten-acre spacing rule. Because the density of the East Texas field held each well's allowable to about 20 barrels, the leaseholder surrounded by small tracts was permitted to produce 480 barrels a day while the unfortunate neighbor got only 80 barrels on the same-sized lease.

Naturally, many large landowners argued that they should not be thus penalized for following the spacing rule, and Commissioners agreed with them. The logical way out of this morass of exceptions was to institute compulsory pooling—that is, to deny small landowners and leaseholders permits to drill and thus compel them to combine tracts to attain a proper acreage for drilling, with an understanding that revenues be shared. Major companies, engineers, and economists constantly urged this course on the Commission. Very occasionally, Commissioners would rule against a request from a small landowner and refuse to grant a first-well Rule 37 exception.

Under these circumstances, the producer would often go to court and succeed in getting an injunction forcing the Commission to grant a permit. Pooling, the rational way out, was thus blocked.

The Commission therefore evolved a policy that went the other way, into even greater complexity. By the late 1930s the "eight times acreage" doctrine was the rule of thumb used by the Commission to determine spacing on large tracts in fields like East Texas that were dominated by small tracts. If an operator asked for a Rule 37 exception, an imaginary circle was drawn around the property, encompassing eight times the area of the lease, and the well density of that circle was calculated. If it were more densely drilled than the operator's lease, that person was granted enough wells to bring the density into equivalence with the surrounding tracts. If the circle were of equal or less density, the request was denied.[20]

As any given operator drilled more wells to bring a tract's density into line with neighboring tracts, the general density of the area increased. Other nearby operators were entitled to more wells under the "8 × A" rule, which created still greater density, which led to more wells, and so on. Ripples of adjustment spread out from every new well in the field, and economic waste proliferated.

As with spacing, so with allocation. The early inclination of the Commissioners was to give each well a set allowable, but the courts insisted that some consideration must be given to the productive potential of the well.[21] Large producers argued for allocation based on acreage or on some similar measure of reserves in place.[22] The more weight an allocation formula gave to individual wells, the more it favored small producers, and the smaller the acreage, the greater the bias. By the mid forties, the Commission had settled on a formula that based one-half of the allocation in oil production on each well and one-half on acreage. Because a typical gas well can drain a much larger area than a normal oil well, the formula for them based one-third of the allowable on the well and two-thirds on acreage.[23] This discriminated severely against the large tracts, but at least it was an objective, consistent policy.

The allocation problem was further complicated by the effort to protect marginal or stripper wells. These are wells on relatively unproductive acreage which, consequently, extract little petroleum. A well drilled to a depth of two thousand feet or less is a stripper if it produces fewer than ten barrels a day, or if it is at a depth of four thousand feet and produces fewer than twenty barrels, or if it is six thousand feet deep and produces fewer than twenty-five barrels.

Because strippers generally cost as much to operate and maintain as do more productive wells, the oil that comes from them is relatively expensive. In order for the owners of these wells to show a profit, the price of oil must be relatively high. Economists argued that strippers were absurdities and should be abandoned, thereby allowing more productive wells to extract more and hence cheaper oil.[24]

But this position struck many Texans (especially those who owned strippers) as the opposite of conservation. In the latter stages of depletion of any field, production falls so that all its wells eventually become marginal. Although the amount of production from these wells in any given year is a minute proportion of the total recovered from the field during its period of flush production, over the long run the recovery totals are still impressive. In some cases, more oil is recovered in the stripper stage than was recovered in a field's flush stage. Moreover, when the amount of production from all marginal wells is added together, it constitutes a sizable percentage of the nation's reserves.

Because the economically efficient course of development envisioned the complete loss of these reserves and because the owners of marginal wells were numerous, vocal, and well organized, the state legislature, the courts, and the Railroad Commission agreed that they must be protected. In 1931 the legislature exempted strippers from prorationing, and the Commission attempted in additional ways to protect them from market forces.[25] Consequently, wells capable of producing great volumes of cheap oil were restricted by allocation formulas, while tens of thousands of marginal wells, each producing a few barrels of expensive oil a day, were allowed to flow unchecked. As a result, by 1941, 31 percent of Texas' 100,050 wells were unprorated strippers and produced 6 percent of the state's total oil.[26] Once again, the large producers fulminated in vain against the Commission's rules.

THE HAWKINS DECISION

They also filed lawsuits but received no more satisfaction. Throughout the thirties and into the forties, major companies challenged in court the whole structure of Railroad Commission decisions with regard to within-field spacing and allocation. Judges attempting both to encourage conservation and to protect the small landholder could no more create a viable policy than could the Commission. In 1938, legal historian Robert Hardwicke counted seventy-five court opinions in Texas spacing cases, none of which had managed to bring clarity or consistency to the regulatory process.[27] Courts often upheld the authority of the Commission to decree spacing and allocation rules but could provide little guidance about how those rules were to be made coherent or fair.

The larger producers gave up after 1946. In that year the Texas Court of Civil Appeals ruled on the fifty-fifty allocation formula promulgated by the Commission for the Hawkins field, also in the Woodbine of East Texas. Not only did the judges uphold the Commission's rules, but they also explicitly stated that the interests of the small landowner were to be protected, whatever the costs to the major companies and the rest of the country:

> . . . the owner of an "involuntarily" segregated tract cannot be denied the right to drill at least one well on his tract however small it may be. From which it would seem that his allowable cannot be cut down to the point where his well would no longer produce, nor below the point where it could not be drilled and operated at a reasonable profit.[28]

With the Hawkins ruling, the major companies conceded defeat and did not challenge a Commission allocation order in court for more than a decade.

CONCLUSION: SPACING, ALLOCATION, AND THE PRICE OF PRORATIONING

By 1946, the political bargain that had won the Railroad Commission the power to prorate had become institutionalized. By holding state production to the market demand, the Commissioners stabilized the industry; by favoring the small producer and landowner in spacing policy, they guaranteed a place for the little guy within that industry. This solicitude for the small producer was so marked that some contemporary observers wondered aloud whether the Commissioners, as the political prisoners of the independents, were incapable of making decisions that would spark the opposition of that segment of the industry.[29]

The critics somewhat exaggerated the clarity of the political coalitions of the time. Certain majors and large landowners exhibited little understanding of the larger consequences of their individual actions; certain independents and small landowners were at the forefront of the fight for more rational production. But, in general, it was accurate to contend that in most policies the Commission was the protector of the industry's little people and that the policies which came from this leaning would not make for efficient development of the state's resources.

In not all issues, however, were the Commissioners incapable of taking a broader view. That they were able to rise above political pressure and adopt a long-run, public interest view was demonstrated on another front. While the Commission dealt with spacing and allocation, it was also wrestling with the problem of the waste of natural gas.

POLICY-MAKING EPISODE NUMBER THREE
Gas Flaring

From the earliest days of oil production, the industry had problems with natural gas. The difficulties all rested, at base, on a fact that seems incredible to our gas-hungry age: the stuff was practically worthless. In 1930, when oil sold for over a dollar a barrel, the price of natural gas was 3.6 cents for a thousand cubic feet.[30] At an equivalency of six thousand cubic feet of gas to one barrel of oil, this meant that oil was five times more valuable than gas in terms of its heating capacity.[31]

But even this low monetary value was mostly theoretical, for it recorded only that gas which found a market. Unlike oil, which can be temporarily stored and easily transported, gas is a difficult substance to handle. It is hard to store and transport, dissipates quickly, and is likely to ignite and explode.[32] Moreover, whereas oil can be used for lubrication and auto fuel as well as for heating, natural gas is not much good for the first two.

As a result, even the low price of gas failed to express its uselessness. When Amarillo, situated next to an ocean of gas in the Panhandle, spent sixty thousand dollars advertising its abundance nationally, it found not a single buyer. The city administration then offered free gas for five years to any industry that would move to Amarillo and employ fifty or more people, still with no takers.[33] This typifies the situation of the petroleum industry in general for its first eighty years. Most producers regarded oil as the only hydrocarbon worth searching for.

THE ECONOMICS OF NATURAL GAS

For the leaseholder who owned a well situated over the 70 percent or so of natural gas that occurs unassociated with oil, the low price was at worst an annoyance. In the early days of the industry, discovery wells in gas fields were often simply capped and forgotten.

There were, however, a few uses for gas. It heated the homes of communities that chanced to be situated near a field or that happened to be at the terminus of one of the then rare gas pipelines. It was used as boiler fuel by the few industrial concerns that were then located in Texas. It could be burned to produce the carbon black that was employed by the rubber industry, and, as the thirties progressed, some companies learned to use it to make other chemicals.[34] The major economic problem caused by gas in the early part of the decade, however, arose from the fact that it could be employed to produce a liquid known as condensate gasoline.

When natural gas is allowed to expand suddenly, as it does when let out of a well bore, somewhat less than 10 percent will condense into a liquid almost indistinguishable from refined gasoline. This process is called stripping gas (no relation to stripper oil wells). Condensate gasoline could be used like the refined variety to power automobiles. Companies could make a profit by setting up over a gas field, marketing the stripped condensate, and simply releasing or venting the remaining nine-tenths into the atmosphere. Early in the history of this practice it was discovered that natural gas in the atmosphere is a deadly hazard, so producers began to run it up pipes and burn it at the top. The flames from these pipes are known as flares. The desire of some producers in the early 1930s to retrieve the 10 percent of stripped condensate and flare the rest presented the Railroad Commission with one more regulatory problem.[35]

Stripping and flaring of unassociated gas clearly presents a waste of a natural resource. To the conservationist, the method of saving such gas is fairly straightforward: require producers to return it to the reservoir after they have removed the condensate. For a second type of gas production, however, the conservation problem is much more complex. This is known as the extraction of casinghead gas.

Gas is always dissolved in oil in underground reservoirs, and it is often found in a gas cap at the top of the oil-bearing rock formation. When the oil is extracted, the dissolved gas is an inevitable by-product. There is no known method of producing oil without simultaneously bringing up large quantities of gas. Historically, about 30 percent of the gas that is produced has been of this associated type.

Within the well, the petroleum travels to the surface inside a metal tubing. The tubing is held in another set of pipes called the casing. At the top of the well, a metal device called the casinghead connects the casing and the tubing. When oil arrives at the mouth of the well, the associated gas dissolved in it escapes into the casing and out of the casinghead. It is therefore called casinghead gas.

For the casinghead gas that was extracted with oil in the 1930s, the prospects for productive employment were even less than for unassociated gas. Its rate of production could not be controlled, for it was an inescapable by-product of oil extraction. Because its supply was dependent on the demand for oil, it was unattractive to gas pipeline companies, who preferred to contract for a steady, predictable supply of unassociated gas. The market, which was weak for unassociated gas, was thus practically nonexistent for casinghead gas. To put it back into the ground was expensive. So, casinghead gas was almost invariably flared.[36]

The combination of the low gas prices, the technology of condensation, and the uncontrollable nature of casinghead gas production caused a waste that staggers today's imagination. There are no reliable figures on the total volume of gas dissipation in the seven decades after Colonel Drake drilled the first oil well in Pennsylvania in 1859, but it must have amounted to trillions of cubic feet, as oil field after oil field was allowed to blow off its entire accumulation of associated gas, and stripping plants in unassociated gas fields utilized only a tiny proportion of the resource.

The waste in Texas in the 1930s and 1940s continued apace, especially with regard to casinghead gas. According to many accounts, motorists could drive for hours at night in parts of Texas in those years and never have to turn on their automobile lights, because the casinghead flares illuminated the countryside. Miles away from any major oil field, newspapers could be read at night by the light of these flares. It was not that people in the industry were insensitive to the loss of this resource, but they saw no alternative to flaring the gas if they wished to produce the oil.

Historian Maurice Cheek estimated that in 1934 roughly a billion cubic feet of unassociated gas was stripped and flared daily in the Panhandle. Assuming that this total was matched by casinghead flares from oil fields (especially the East Texas field), then in the neighborhood of three-quarters of a trillion cubic feet was lost that year. The best estimate from the early 1940s is that one and a half billion cubic feet of casinghead gas was flared each day from the state's larger fields; that would make the state total for all fields about two and a half billion a day or over nine-tenths of a trillion a year.[37] As the yearly consumption of natural gas in the entire United States was only about twenty trillion cubic feet in the mid 1970s,[38] a sizable proportion of the nation's potential energy supply vanished in a glow of prosperity in Texas before 1949, when gas flaring was finally controlled.

The simple fact of the destruction of a natural resource was an obvious reason for many people to advocate the elimination of gas flaring. There was another, more subtle reason, however, advanced by engineers and others of scientific training: they wished to save casinghead gas in order to increase the state's recoverable oil reserves.

In most reservoirs, the propulsive force that moves the oil to the well bore is provided by gas pressure. The higher the pressure, and the longer it lasts, the more oil is ultimately recoverable. The faster the pressure is depleted, the smaller a proportion of the oil in place can be brought to the surface, without using sophisticated and expensive recovery techniques.

When a method of returning casinghead gas to the producing rock formation was invented in the early 1930s, the life of oil fields could be greatly extended. But this pressure maintenance was costly. Although in the long run it might quadruple the amount of oil extracted from a field, in the short run it cost the producers money to no obvious advantage. Their perspectives dominated by short-run profit considerations, oil producers resisted suggestions that they should take the necessary steps to preserve gas pressure.

As the agency charged with conserving the state's oil and gas resources, the Railroad Commission was at the center of the controversy over gas flaring. Throughout the 1930s and 1940s, the various Commissioners and their staff attempted to deal with the technological and political problems of gas conservation. This conflict placed the Commissioners in a difficult position. To have forced "uneconomical" conservation on the industry would have incited it to strong opposition. As politicians subject to electoral defeat, Commissioners were not eager to provoke hostility from this, their basic constituency. Until 1947, consequently, Commissioners treated the problem of gas gingerly. The Commission's engineers, however, worked for two decades to eliminate the waste of gas and finally created a political momentum that succeeded in making the agency move decisively to stop the flaring.

WHAT IS AN OIL WELL?

In 1899, the state legislature had passed a comprehensive conservation law, later amended several times, in which the flaring of unassociated gas from a gas well was prohibited. In 1925, the legislature passed a law permitting the flaring of associated gas from an oil well. In theory, the Railroad Commission's task in enforcing these statutes was easy: it should prohibit the flaring of gas from gas wells but not from oil wells. In practice, however, the two laws created a regulatory nightmare.[39]

Even in a gas field, there may be traces of oil. Gas that is unassociated with oil in a practical sense may be associated with minute quantities in a technical sense. Furthermore, the amount of oil found in a gas field may vary widely. This means that a regulatory agency must be able to answer the question: how much oil can be found in a well before that well stops being a gas well and becomes an oil well?

Moreover, gas expands, contracts, and changes its physical composition under differing pressures and temperatures; under some conditions gas will change into a liquid and vice versa. These circumstances add to the difficulty of determining whether it is a gas or an oil well that is under consideration. With producers permitted to flare from one type of well but not from the other, this technical problem of differentiating gas from oil wells became a political problem.

Texas statutes defined a gas well as one that produced 100,000 or more cubic feet of gas for every barrel of oil. If the ratio were less than 100,000 to 1, the well was classified as oil, and the gas was casinghead and could be flared. If the ratio were higher, the well was classified as gas, and its product could not be flared. It was thus in the interests of operators who owned wells that produced both gas and oil to have as many of them as possible classified as oil, so that they could flare unmolested. They played cat and mouse with the Commission's overworked inspection staff to attain that goal.[40]

One of the critical junctures in this struggle occurred in 1934. A number of the fields of the time produced large quantities of a clear petroleum liquid known as water-white oil along with conventional oil and gas. Their operators treated this liquid as though it were oil. By doing so, they made the gas-oil ratio of their wells 40,000 to 50,000 to 1, thereby causing them to be categorized as oil wells. This permitted the operators to flare a large amount of gas after they had retrieved a small amount of oil. If the water-white oil were to be classified as gas, however, the ratio of the wells in question would rise above the 100,000 to 1 threshold, the wells would be reclassified as gas, and the flaring could be halted. The Railroad Commission hired Eugene P. Schoch, a chemist at the University of Texas, to investigate this water-white oil.

Schoch instructed one of his students, Jack K. Baumel—later to be chief engineer for the Commission—to take samples of water-white oil from the Agua Dulce field near Corpus Christi. Back at the University of Texas laboratory, Schoch and Baumel put this liquid into a high-pressure separator, which allowed them to recreate reservoir conditions by raising the temperature to 247 degrees and the pressure to 3,700 pounds per square inch. When they did this, the water-white oil turned into a gas. The substance had been gas in the reservoir but had turned into a liquid upon reaching the surface.

With this evidence, the Commission's engineering staff recommended that hundreds of oil wells be reclassified as gas and their operators ordered to stop flaring. The operators naturally objected to this order, but, in a series of judicial opinions known collectively as the Clymore case, the Texas courts backed the Commission to the hilt.[41]

Now the operators were in difficult circumstances, for they had to find something to do with the gas produced with the oil or shut down their wells. The Commission engineers suggested that they could keep their oil wells flowing and make further revenues from gas by using a process called cycling. The Schoch-Baumel experiments had demonstrated that the gas in many fields like Agua Dulce was rich in condensable liquids. Why not, suggested the engineers, collect the (wet) gas on the surface, extract the liquids, return the remainder of the gas (now dry) to the reservoir, where its added pressure could aid in the production of more oil and wet gas, and so on? The condensate from the gas could be marketed, as could, of course, the oil.

If they wanted to continue production, the operators had no choice but to begin cycling. The first commercial cycling plant began operating in the Cayuga field of Northeast Texas in March 1938. To everyone's relief, it proved profitable. By 1942 there were over twenty-nine cycling plants in Texas, processing over forty-four billion cubic feet of gas a year. By losing a technical battle to the Railroad Commission, petroleum producers had improved their economic position.[42]

One of the conflicts over gas flaring thus ended happily for all concerned. With superior technical imagination and a determination to take the long view, the Railroad Commission had finessed one of the problems of gas conservation. Other struggles, however, were more complicated.

PROBLEMS IN THE PANHANDLE

The Panhandle field was to gas what the East Texas field was to oil. One hundred twenty-five miles in length and an average of twenty-five miles wide, when discovered in 1918 it is estimated to have contained between fifteen and twenty-five trillion cubic feet of natural gas.[43] Despite the 1899 conservation law, in the 1930s some operators began to attempt to acquire the right to strip and flare this gas. For the first five years of the decade, while the oil proration war raged in East Texas, a lower-keyed but no less complex struggle occurred over the conservation of gas in the Panhandle. As with the oil proration episode, the forces of conservation in this encounter suffered initial defeat. In 1933, under lobbying pressure from stripping interests, the legislature passed a law specifically permitting Panhandle operators to strip and flare gas.[44]

The enormous volume of gas blown into the air under this law—over a billion cubic feet a day—caused both a public outcry and a lobbying campaign by pipeline companies (who themselves wanted the gas) to persuade the legislature to repeal the 1933 law. This effort succeeded in 1935 when House Bill 266 forbade the production of gas in any manner that caused underground waste (for example, the stripping and flaring of unassociated gas) and set up a system of gas prorationing. The Railroad Commission was empowered to enforce this act.[45]

Most portions of House Bill 266 were upheld in the courts, and the Commission swiftly put an end to stripping and flaring operations in the Panhandle. The efforts of the Commission to prorate gas production in the Panhandle, however, were invalidated by a federal court. The United States Supreme Court upheld the federal court but left the Commission in doubt as to just what its powers were in the area of gas prorationing.[46] Because of these decisions, in the following decade Commissioners were uncertain about the extent of their authority over gas production.

PROBLEMS WITH CASINGHEAD GAS

Despite the Commission's difficulties with prorationing, the passage of House Bill 266 in 1935 effectively ended the problem of the flaring of unassociated gas. The waste of casinghead gas, however, increased during this period. Throughout the decade, the Texas industry continued to discover giant fields: Conroe in 1931, Tom O'Connor in 1934, Wasson in 1936, Levelland in 1938, Hawkins in 1940, and so on.[47] Every new field opened to development meant another great volume of casinghead gas doomed to vanish unproductively. As the problem of the flaring of unassociated gas declined in importance, therefore, the magnitude of lost associated gas grew ever greater.

Without actually forbidding the flaring of casinghead gas, Commissioners of the late 1930s and early 1940s attempted to encourage the conservation of this resource. For example, they issued a statewide order establishing a permissible gas-oil ratio of two thousand cubic feet per barrel. If the flared gas amounted to more than two thousand cubic feet for every barrel of oil produced, the well was subject to being declared inefficient and might, after due hearing, have its oil allowable reduced.[48] This and other Commission efforts at conservation were well intended, but they made little impact on the problem of casinghead gas, for two reasons.

First, the Commission had such a small staff that it could not enforce the order. In the case of oil production, there was an actual substance flowing in commerce, which could be monitored. In the case of gas, however, the substance in question was destroyed as soon as it was produced; there was no commerce to monitor. The only way to enforce the order was to keep testing the gas-oil ratios of many hundred thousand wells. The Commissioners did not have anything like a staff adequate to the task, although they did what they could with the staff at their disposal.

Second, even if there had been some means of ensuring that all oil wells were kept within the 2,000 to 1 boundaries, only the most wasteful wells would have been eliminated. The basic problem of the wanton flaring of "useless" gas was not touched by creating "acceptable" levels of destruction. Within the prescribed limits or not, the flared gas was gone forever. The only way to save this resource was to outlaw the flaring completely.

Fearing to stir up opposition from the industry, and reluctant to impose economic hardships on it, the Commissioners of the 1930s and early 1940s refused to take this step. They were happy to encourage conservation by approving schemes for repressuring fields with gas,[49] but they would not consider compelling the industry to act responsibly. It would take the elevation of a petroleum engineer to the Commission to change this attitude.

THE GAS WAR

In 1939, a young petroleum engineer named William Murray had taken a crew to South and West Texas, testing gas-oil ratios for the Railroad Commission. In the course of observing thousands of wells and testing several hundred, he had been appalled at the enormous volume of gas that was being lost through casinghead flares. Most of this flared gas did not show up on Commission reports, for the operators were lax about keeping records, and the Commission did not have the staff to police them. There was little that Murray could do to stop the waste, since most of the ratios were within the acceptable limits, but he nevertheless resolved to do something about it in the unlikely event that he ever got the chance. Murray worked for the Commission another two years, then left and was employed by the Federal Petroleum Administration for War until 1943. He then resigned and returned to Texas, where he went to work for private industry.

Because of the extreme importance of petroleum in fighting the war, some officials in the federal government had by this time grown concerned about reports of gas waste in Texas, and the Federal Power Commission was known to be considering imposing its own regulatory authority on the state industry. Sitting on the Railroad Commission with Ernest Thompson at this point were Olin Culberson and Beauford Jester, all of them vociferous advocates of states' rights. They attempted to forestall federal meddling in Texas conservation policy by convincing the FPC that the state had the gas problem under control.

In December of 1944 the Commission scheduled a special hearing to discuss the gas-flaring problem, among other topics. The official figures showed that over 400,000,000,000 cubic feet of casinghead gas had been produced in the state in 1943, of which only 3,690,787,000 cubic feet, or less than one percent, had been flared.[50] Ernest Thompson argued that this volume was reasonable and posed no threat to conservation. But William Murray, attending the hearing as a private citizen, stood up and stated that he knew from personal experience that the received figures were gross underestimations, that from ten to twenty-five times the official estimates was being lost, and that this was indeed a serious conservation problem.[51]

This accusation drew some press coverage and embarrassed the Commissioners. Gas flaring is the sort of dramatic issue that is news, for the burning of a natural resource is easily understood by a mass audience. So the Commission was forced to make at least symbolic motions to solve the problem. The Commissioners appointed an industry-wide committee to look into the flaring problem and asked Murray to chair it.[52] But, since Murray knew that some of the worst offenders were on the committee, he declined and asked instead to chair a smaller group, composed only of engineers, that would report to the larger body.

The Murray Committee report, released in November of 1945, caused a furor. Many of the most prominent men in Texas were seen to be contributing to a waste of casinghead gas of almost a billion and a half cubic feet a day or 57 percent of the state's total production.[53] The big producers complained vigorously in private about this engineers' committee to their friends on the Commission. Publicly, they argued that an order to stop flaring the gas would ruin them.[54] But, because of the publicity created by this very forthright report, it would have been dangerous for the Commissioners to ignore it. Instead, they stalled.

Meanwhile, some members of the industry were awakening to the magnitude of the problem. The Murray Committee report, by compiling accurate figures on the volume of gas lost, forced the industry to confront its own profligacy, and those producers who were more farseeing became convinced that this resource must be conserved. In particular, Dan Moran, president of Conoco and an active member of the industry committee, concluded that the flaring had to be stopped and vigorously supported Murray and his fellow engineers within the industry.

While the struggle over flaring gas was taking place, the petroleum industry was evolving in a direction that favored the conservationists. During the war, technical advances had been made in the process of cycling, and it became more feasible to employ that process with casinghead gas.[55] Also during the war, the Big Inch and Little Inch oil pipelines had been built from Texas to the Northeast, to avoid the attacks of German submarines on tankers. With the war over and oil moving again by sea, the pipelines were capable of carrying gas, and people in the industry began to discover that they could sell it once they had the means to move it. Finally, the war had given a great boost to the petrochemical industry, to which natural gas was becoming important.[56]

As a result of all these changes, the price of natural gas began to rise. In 1940, it had been only 1.8 cents per thousand cubic feet, but by 1947 it was up to 3.7 cents.[57] This was not enough to make conservation profitable for most operators, but it did tend to forestall panic in the industry at the thought of having to eliminate flaring.

Moreover, the most dreaded of all Texas nightmares, federal control, threatened to overtake the industry if something were not done about the gas situation. In 1946, the Federal Power Commission held a series of hearings on gas waste, with the obvious implication that it might resolve to extend its regulations over the state industry for reasons of conservation. At a hearing in February in Houston, six Texas officials, including two Commissioners, told the FPC that they were making great progress in eliminating waste and that they didn't need federal help. If the Murray report were to be believed, however, they *weren't* making progress, and that knowledge made figures in both public and private life in Texas nervous.[58]

All these forces would have combined, sooner or later, to stop gas flaring. But, in January of 1947, a political act occurred that made it sooner. Commissioner Jester had been elected governor in 1946. One of his first acts as chief executive was to appoint William Murray, the crusading engineer, to serve out his own unexpired term on the Commission.

As Murray had spent the previous two years disrupting the most important industry in Texas, the governor's choice may seem surprising. But Jester had stated that he thought that the most important problem facing the industry was gas flaring and that Murray was the best man to handle the problem; there is no reason to disbelieve him.[59] Jester must be given credit for possessing considerable political courage.

The confirmation of the appointment could have been stopped in the state senate, for of course there was much opposition to Murray within the industry. But that opposition could have succeeded in blocking Murray only if it had represented the unanimous sentiments of Texas producers. A small group of very active, influential independents, including Robert Foree and Glenn McCarthy, backed Murray vigorously and prevented any movement to contest the appointment. He was confirmed easily.

With Murray's ascension, the tone of Railroad Commission activity altered abruptly. Because they had been worried about imposing economic burdens on the industry, Culberson and Thompson had been at best halfhearted in their efforts to stop flaring. The arrival of a colleague committed to eliminating the waste, however, coinciding with rising gas prices and a nosy FPC, made it clear that they would have to move or lose their position as leaders. The course of political expediency suddenly coincided with the path of political virtue. And so, faced with the inevitable, Thompson and Culberson jumped on the gas bandwagon with vigor. The Railroad Commission became a conservation tiger.

On March 17, 1947, the Commission issued an order shutting down all 615 oil wells in the new Seeligson field in South Texas until a cycling and compression plant was completed and the flaring of casinghead gas was eliminated. To say the least, this order shocked the industry.[60] Shell, Sun, and Magnolia (now Mobil), big operators in the field, immediately filed suit. Former governor Dan Moody, the attorney for Shell, was confident of victory.[61] But the Texas Supreme Court upheld the order.[62]

Having won its test case, the Commission proceeded to issue a series of orders shutting down seventeen fields for flaring. These orders were also appealed in the courts, but by 1949 the Commission's power to protect gas had been solidly upheld. In the Flour Bluff case, the court made it clear that it would do the industry no good to plead that saving gas was unprofitable:

> If the prevention of waste of natural resources such as gas is to await the time when direct and immediate profits can be realized from the operation, there would have been little need for the people of Texas to have amended their Constitution by declaring that the preservation and conservation of natural resources of the state are public rights and duties and directing that the legislature pass such laws as may be appropriate thereto . . . for private enterprise would not need the compulsion of law to conserve these resources if the practice were financially profitable.[63]

There was more legal and political maneuvering in the coming years, but it was just skirmishing.[64] The war had been won by 1949, with the Railroad Commission the unquestioned victor. Henceforward, with relatively insignificant exceptions, casinghead gas would go into pipelines or back into the ground.

THE ANATOMY OF STATESMANSHIP

The elimination of gas flaring in Texas must be counted as one of the great victories for conservation in the history of the United States.[65] It extended the lives of oil fields, saved the nation tens of millions of dollars, and postponed the day when natural gas would have to be imported. It is not just the value of the conservation that is important, however, but the fact that it was imposed against the nearly unanimous opinion of the petroleum industry. Regulatory agencies, state or federal, are not known for their willingness to impose economic burdens on the interests they regulate. Yet, by decreeing that gas must be conserved, the Railroad Commission was costing the industry money, for in their first years many of the cycling and compression plants that had to be built to process casinghead gas ran in the red.[66] The gas-flaring episode is a clear example of the Commission taking a politically dangerous course and forcing the industry to act in the public interests. What were the components of this policy?

First, the industry was not completely united in opposing the Commission's decisions, because the improving market for gas prevented producers from anticipating economic disaster if flaring were prohibited. Before its first Seeligson order, the Commission held hearings in Corpus Christi, Houston, Midland, and Fort Worth on the gas situation. Commission engineers testified at these hearings that, at the current price of gas, a cycling plant could pay for itself in just over two years. Such reassurances kept the resentment at the flaring orders from causing a mutiny.[67]

Second, the industry was in a ticklish position in trying to oppose the Railroad Commission. Majors were not about to upset the prorationing applecart by striking at the agency's authority, and the independents were not eager to disrupt a body that favored them in its spacing and allocation decisions. Furthermore, Ernest Thompson, as the hero of the crisis of the early 1930s and as the dominant figure on the Commission in succeeding years, had by 1947 acquired a huge amount of prestige within the industry. When he moved to eliminate flaring, it was unthinkable that the industry would oppose him politically; it must content itself with legal challenges. The Commission was in a relatively secure position, despite the fact that it was imposing a hardship on its own constituency.

Third, as was uncommon in regulation controversies, the courts backed the Commission vigorously and consistently. There were no ambiguities, reverses, qualifications, or self-contradictions in the three important flaring cases of 1947 and 1949, as there had been in the prorationing and spacing decisions of the previous two decades. With internal and external political circumstances favoring the Commission, judicial approval made its flaring orders irresistible.

Fourth, the presence on the Commission of an engineer dedicated to the elimination of flaring made for a pressure that was impossible for the other Commissioners to resist. As will be shown in chapter 4, Murray's technical approach to oil and gas regulation could be, and often was, blocked by Culberson and Thompson on other subjects. In combination with the variety of circumstances surrounding the controversy over flaring, however, his insistence on conservation in that one issue could not be resisted. The final irony of the regulatory battles of this era was that the Commissioner most responsible for defending the public interests was the one who had the least political ambition and, hence, did not worry about offending the industry.

CONCLUSION

Commissioners handled the two important controversies that occurred between 1935 and 1950 in quite different ways. They dealt with the spacing and allocation problem by siding with one segment of the industry against the other, sacrificing economic efficiency and the interests of the large producers for political support and the interests of the industry's little people. The flaring problem could not be managed this way, because the conflict was not between parts of the industry but between the industry and the rest of society. Commissioners handled this problem first by ignoring it, then by assuming the mantle of tribune of the people and forcing conservation on a reluctant industry.

In each struggle, the choices of the Commission were dictated partly by political circumstances and partly by the personalities of individual Commissioners. The result of their choices was that, by

1950, they had mastered the Texas industry and created a definite system within which all producers and royalty owners had to operate. In the following decade, the virtues and vices of that system began to be realized.

4. The Balance Wheel, 1950–1965

As Texas goes, so goes the industry.—Folk saying, oil industry, 1950s

The decade of the 1950s was a great one for the Railroad Commission. Texas was preeminent in the world of oil, and the Commissioners dominated the state industry. In mid decade, 43 percent of Texas' land was either producing petroleum or was under lease.[1] In 1953, the state's production of 2.75 million barrels of oil a day from its more than six thousand fields was 45 percent of the United States total, twice the production of the Soviet Union, and more than the entire output of the Middle East.[2]

The Commissioners were acknowledged to be the dominant force of this industry. As the decade began, Ernest Thompson possessed eighteen years of regulatory experience and Olin Culberson nine. While William Murray had been on the Commission only since 1947, his expertise as a trained petroleum engineer made up for any lack of experience. To the prestige the Commission had gained by having mastered the proration problem fifteen years before had recently been added the respect it had acquired by ending most gas flaring. In the 1950s, industry and academic observers alike thought of the Railroad Commission as the most important single component of the domestic petroleum industry.[3] Moreover, there was no turnover on the Commission during the 1950s. Thompson, Culberson, and Murray served the whole decade, which merely enhanced the Commission's position as the legitimate tribunal of the industry.

The complete authority wielded by the Commission in those years seemed to be personified by Ernest Thompson. Thompson made himself both the arbiter of individual differences within the Texas industry (see chapter 7) and the spokesman for the entire range of domestic producers. He made scores of speeches during the decade, not only to industry gatherings but to military groups, press associations, congressional committees, and, in general, anyone who would listen. His message varied with the audience and the current issue, but its underlying theme was consistent: a strong domestic petroleum industry was a good thing for the United States, and the Texas Railroad Commission was doing all it could to nurture that industry. The notions familiar in the 1970s and 1980s that there might be fundamental conflicts between consumers and producers, or between Texas and the rest of the country, were completely absent from Thompson's speeches and testimony of the 1950s. He consistently argued that a healthy oil and gas industry was in the best interests of everyone.[4]

Olin Culberson and William Murray were not publicly silent in the 1950s, but their statements tended to be addressed to audiences within the industry. They left the banner carrying to the more colorful Thompson. And, for their part, people in the industry recognized Thompson as one of their most effective representatives and honored him every way they could. In 1951 he became only the sixth man to be awarded the American Petroleum Institute's Gold Medal for Distinguished Achievement.[5] In 1954, a laudatory biography by Texas' best-known oil writer was published.[6] Throughout the decade, members of the industry staged testimonial dinners for him. Thompson, and the Railroad Commission, seemed to be at the pinnacle.

The mechanics of the control system that enabled the Commission to exert such power were relatively simple and informal. This system rested on two types of hearing: field hearings, at which Commissioners gathered the information to enable them to set the spacing and allocation rules for a given field, and statewide hearings, at which the Commission would decide the overall level of Texas production for the coming month. In neither of these did the Commissioners stand on procedure.

Field hearings were relaxed and friendly, and transcripts were only rarely kept. Commissioners maintained good personal relations with the lawyers and engineers who appeared before them (see chapter 7), and their discussions, while rigorous, were informal and lively. The atmosphere of field hearings more closely resembled that of a family gathering than of the quasi-judicial deliberation that is the textbook model of regulatory procedure. The give-and-take between Commissioners, their staff, and the lawyers and expert witnesses for the companies was vigorous and thorough but sociable.[7]

Statewide proration hearings were somewhat more structured. Each month, representatives of the important oil-purchasing companies (mostly majors) would gather in Austin. Each company had a certain quota of crude oil that it wanted to buy, based on the amount it thought it could retail in the coming month. But the actual amount it requested was more a collective than an individual decision. Because of the pattern of across-the-board statewide prorationing that the Commission had enforced since 1947, the company representatives (who called themselves crude-oil men) always had to make some cooperative adjustments in the amount they wished to buy.

Humble, for example, might find that it could fill its needs in a given month with a 50 percent allowable, and Gulf might discover that it required only a 30 percent allowable. Across-the-board prorationing prevented them from requesting different allowables. So, the night before the hearing, the crude-oil men from each company would meet and agree upon a joint request of, say, a 40 percent allowable. Gulf would consent to permit Humble to purchase an extra 10 percent of crude oil from Gulf's theoretical allowable in the coming month.

The next month, the position of each company might be reversed, and Gulf might need the larger allowable. In that case, Humble would agree to adjust its own request to aid Gulf, and so on. This swapping of crude oil, which took place among the major purchasers on a large scale before each statewide hearing, prevented the Commission's across-the-board proration formula from causing dislocations in the industry.

The next morning, the crude-oil men would present their nominations for the amount their company wished to buy. The Commissioners would consider these figures, along with the United States Bureau of Mines estimates of national market demand for the next month; other factors such as industrial activity and the weather were also considered. The main indicator was the quantity of and change in stocks of four principal products: heating oil, gasoline, distillate, and crude oil. From these figures the Commissioners would estimate the market demand for Texas crude for the coming month. The statewide allowable was usually somewhat below the purchasing companies' nominations, but, because of the swapping process the night before, they were easily able to adjust their expectations.

As the statewide allowable was always lower than the state's potential production, the Commissioners would proceed to prorate. If, as was common in the 1950s, they determined that only one-half of the state's maximum production would satisfy the market demand, they would announce a statewide allowable for the next month of fifteen days. This meant that for only half of the month could any given well be operated. The operators all knew the theoretical maximum of their wells—most calculated from the 1947 yardstick, a few, as in East Texas, from special field rules. Knowing the maximum permitted rate of flow from their wells, and knowing the proportion of that maximum that the Commission had decreed as acceptable for the coming month, operators could easily calculate how much oil each well would be allowed to produce. The Commission kept records in Austin on each well's production and monitored them to ensure that no hot oil was produced in excess of the monthly allowable.[8]

Under the 1935 Connally Hot Oil Act, the federal government prevented crude that was produced in violation of state regulations from traveling in interstate commerce. Theoretically, California and Illinois, as states that did not prorate to market demand, could have destroyed state production control by drowning the market in a flood of unprorated oil. But California, though a major producing state, was isolated in its own market area, and production in Illinois was both small and short-lived.[9] The states that practiced prorationing were therefore able to balance supply and demand for the whole country east of California.

As by far the largest producing state, Texas was easily the most important force in this system of national prorationing. As other states rose or declined in output, the Commission would adjust Texas' proration schedule to make the sum of the national production equal the national market demand. It was thus the balance wheel for the entire domestic control apparatus and was recognized as such by many observers.[10]

Despite its preeminence during the 1950s, however, the Railroad Commission was not without problems. Two potential calamities that had developed quietly during the decade suddenly hit the Commission with a dual crisis in the spring of 1961. They would result in the destruction of the Commission's spacing and allocation policy and the worst scandal in its history.

POLICY-MAKING EPISODE NUMBER FOUR
Spacing and Allocation, Round Two

After the Hawkins court decision mandating profitable allowables for small tracts in 1946, the Commission's policy of looking after the little guy had become institutionalized. When the operators of a newly discovered field applied for field rules, the Commission's staff examiner would compile the relevant engineering data, often running to several pages, and offer a recommendation. If the field contained small tracts, the examiner would be sure to mention that fact several times in his memorandum. The Commission would then issue rules, ostensibly based on the engineering data but actually based on the presence or absence of small tracts.

If there were no small tracts, the Commission usually followed the recommendations of the field's operators, ordering wide spacing as well as allocation based on acreage or, beginning in 1959, on acreage and the thickness of the producing strata underlying the tract.[11] If the field contained small tracts, however, the Commission issued rules that ensured the smaller operators and landowners a profit. No small producer was kept from drilling a first well. Of 1,538 Rule 37 exception requests in 1956, for example, the Commission denied only 19.[12] If the field were oil, allocation was based half on acreage and half on the well; if gas, the formula was two-thirds to one-third. This ensured that wells on small tracts would drain their larger neighbors.

Big producers always attempted to induce the small producers to join pooling agreements, in which only a few wells would be drilled and the profits would be shared. But, because the Commission's allocation formulas discriminated in favor of smaller producers, they had no incentive to join a pooling arrangement—quite the contrary. As a result, although the Commission continually altered its basic placement policy to make for wider and wider spacing, in any field with small tracts narrowly spaced wells proliferated. Despite the fact that state policy called for forty-acre spacing, therefore, 76 percent of the wells drilled in 1958 were on twenty-acre or narrower spacing.[13]

The attitudes of the three Commissioners to this policy were by no means identical. Olin Culberson, the partisan of the small producer, supported it enthusiastically. Ernest Thompson was willing to go along with it. William Murray, his engineer's sensibilities affronted by its irrationality, opposed it and would have changed it if possible. He occasionally registered his dissent by writing on allocation orders that he was signing his name "because of established Commission policy . . . adopted by a majority of the Commission."[14] Given the political power of the small producers, however, it is doubtful that the policy would have been different even if other personalities had occupied Culberson's and Thompson's positions.

IMPORTS AND OPPOSITION

Nevertheless, external forces were building up a strong opposition to narrow spacing within the industry. For most of the twentieth century, major companies had brought foreign crude oil to American refineries. The volume had not at first been enough to compete seriously with American producers. In the late 1930s and throughout the 1940s, however, majors made huge strikes in the Middle East, and their imports of foreign oil began to increase dramatically as the 1950s began.

The key to imports was their inexpensive production. Foreign oil was so easy to extract that major companies could produce a barrel from a field in the Middle East or South America and ship it to Houston to be refined more cheaply than they could produce a barrel from the fields immediately adjacent to Houston. By 1960, it cost roughly $.20 to produce a barrel of oil in the Middle East, $.80 in Venezuela, and $1.75 in the United States.[15] Some fields were even less expensive; in its huge Ghawar field in Saudi Arabia, for instance, Aramco, the majors' consortium, could probably produce a barrel for no more than $.05.[16] Because they controlled Middle East production through this consortium, the majors set the wholesale prices. They chose to list all crude at U.S. Gulf coast prices and, consequently, they made a high profit on each barrel of imported oil.[17] Naturally, given the glittering profit potential, the majors attempted to bring in as much foreign oil as possible. By 1958, 12 percent of the total crude processed in American refineries was of foreign origin.[18]

But every barrel of outside oil brought into the United States was one less barrel that could be produced domestically. As imports rose, the states that prorated had to suppress domestic production more and more to avoid overburdening the market demand, and thus forcing down prices. By the end of the decade, the Railroad Commission was allowing the wells of Texas to flow at less than a third of their theoretical limit of production.[19]

The majors did not mind this, for they made their big profits on foreign oil. But a low market demand threatened the domestic independents with extinction. If they could not run their wells, they could not make a living. In the latter half of the 1950s, therefore, independents began to hurt, and by the early years of the next decade they were desperate.[20]

Thus arose a political struggle between the independents and the majors that was just as bitter, and lasted longer, than the prorationing battle of a quarter century earlier. Unlike the prorationing contest, this political war was largely waged in Washington. Independents, arguing that the domestic industry had to be protected, urged the imposition of import quotas. Majors (and many others) countered that low oil prices were a boon to consumers and that, therefore, imports should be unrestricted.[21]

Although they were only on the fringes of this conflict, there was never any doubt about the sympathies of the Railroad Commissioners. Beginning in the 1950s and continuing into the 1970s, Commissioners repeatedly insisted, in Texas and Washington, that a healthy domestic industry was essential to national security and that therefore import restrictions should be enforced at the federal level.[22] Moreover, they harassed the majors' crude-oil men at monthly nomination hearings with detailed questions about the quantity and destination of imported oil. Commissioners did not expect that these public displays of hostility would make the majors alter their import policies. As symbolic exercises, however, they served to reassure the independents about the loyalties of the Railroad Commission.

But expressions of support were not enough. When American oil was the only game in town, its price was basically irrelevant, and people in the industry didn't care much about costs. But, with cheap crude flowing in from abroad and strangling domestic production, industry leaders began to look for ways to increase their economic efficiency and thereby lower the price of their product.[23]

Narrow-spacing formulas and stripper wells, as high-cost production units, thus began to come under heavy attack in the late 1950s. The battle over strippers was relatively less violent and will not be discussed here. But the small-tract problem was the object of repeated reform efforts by economists and larger producers, who were increasingly concerned about the health of the American petroleum industry.[24] In 1958, the Oil Industry Conservation Forum, composed of representatives from both independents and majors, was formed to furnish free speakers advocating wide spacing.[25] Articles and editorials appeared in oil publications, urging Texas and Louisiana to modify their spacing policies.[26] Two prominent independents, geologist Michel Halbouty and Johnny Mitchell, launched personal campaigns to persuade their brethren to accept pooling as an alternative to individual wells.[27]

The arguments against small tracts pervaded Texas in the late 1950s and early 1960s. No one who read oil and gas newspaper columns or journals, or who talked to members of the industry, could escape them. As the plight of the domestic producers became worse, the arguments must have been more and more persuasive. In the absence of surveys, industry opinion cannot be measured, but it seems plausible that there was a swing away from toleration of economically wasteful allocation and spacing policies during those years. Interested nonindustry observers, such as judges and state legislators, would have been sensitive to such swings in opinion. But, whatever the general sentiments, the specific interests of those who depended on Rule 37 exceptions and favorable allocation formulas were unchanged. Whatever the general interests, these producers continued to advocate their personal exemptions with vehemence.

COUNTERATTACK

What one writer later termed "Texas' running war over small tract drilling" became as heated in the late 1950s as it had been prior to the Hawkins decision.[28] To ask small producers and landowners to give up their individual wells was to ask them, in effect, to abandon the hope of economic independence. Also, in the absence of a state pooling law, small producers forced to submit to wide spacing would be at the mercy of whatever deals the large producers in a field offered them. Moreover, Texas' drilling industry would be severely damaged by a reduction in the number of wells, a fact that was not lost on drillers.

The smaller independents, smaller landowners, and drilling companies consequently kept up a constant stream of pressure both within the industry and aimed at the Commission, in order to remind everyone of their vulnerability and their determination to fight for what they believed to be their survival. Halbouty and Mitchell were vilified as "tools of the majors," perhaps the worst epithet that one independent can hurl at another.[29] One driller actually threatened Halbouty publicly with bodily harm because, he said, the geologist was taking the food from his family's mouths by advocating pooling.

The attitudes of the Commissioners to this issue are nicely illustrated by the reply of each to a letter Johnny Mitchell sent them in July 1959. Mitchell made it clear that, although he opposed their policies, he blamed the small-tract problem on the producers, not on the Commission. Further confessing that he never expected the campaign against narrow spacing to be won in Texas, he requested an individual reply from each Commissioner.

William Murray responded mildly and obliquely, suggesting

that Mitchell already knew his position and expressing general support. Ernest Thompson was polite but firm in maintaining that the matter was out of the Commissioners' hands, saying, "It is the law that requires us to grant permits on small tracts. The courts have held we must give permits for at least one well on separate tracts." Olin Culberson was, characteristically, more blunt:

> I thoroughly agree with your statement that you and others are fighting a losing fight on wide spacing and you will never win it, regardless of the fact that the majors are behind it and using you boys who *used to be* independent until your interests began to run parallel with that of the majors, until the legal concept of property rights in Texas is changed.[30]

It seems clear that, with this balance of forces on the Commission, a more efficient spacing and allocation formula would not have evolved on its own.

JUDICIAL REVERSAL

In September of 1957, the Commission issued field rules for the Normanna gas field in Bee County near Corpus Christi. As was customary, the examiner's memo to the Commissioners pointed out that the field contained numerous small tracts. It also presented one applicant's calculation that the normal allocation formula would allow these tracts to drain from 450 to 500 times the gas in place under their own plots and, therefore, to "confiscate" their neighbors' property. On the basis of these figures, representatives of Atlantic Richfield (now ARCO), the field's largest producer, asked the Commission to institute allocation based on 100 percent acreage. But one townsite owner "representing himself and his mother who owned royalty under a 28-acre tract expressed the opinion that a straight acreage allocation formula would be discriminating against the small tracts, and asked for some well factor to be included." Several other small landowners concurred. The Commission followed its usual procedure, granting wide spacing but instituting a one-third per-well allocation factor and thus opening the door to a host of Rule 37 requests.[31]

Atlantic Richfield promptly sued, charging that its gas was being confiscated under the Commission's allocation formula. Two years later, while the Normanna case was wending its way through the courts, the Commission approved field rules for another gas field at Port Acres in the extreme southeast corner of the state. Once again, the field's large producers recommended an allocation formula based on a measure of reserves in place (in this case, on acre-feet); once again, small landowners protested; and, once again, the Commission instituted the one-third to two-thirds formula.[32] It was calculable that this formula, given the known qualities of the field, would allow twenty Rule 37 wells, having .65 percent of the total acreage, to drain 14.6 percent of the gas in the field.[33] This time, it was Michel Halbouty who sued the Commission.

Olin Culberson died in June of 1961. Governor Price Daniel appointed former Lieutenant Governor Ben Ramsey to succeed him. Whether or not Ramsey's addition to the Commission would have altered the agency's policy will never be known, for, in March of the same year, the Supreme Court of Texas had overturned the Hawkins decision and declared the Normanna allocation rules invalid.[34] Then, in February 1962, the court invalidated the Port Acres formula and ordered the Railroad Commission to devise some means of allocating production that did not permit small producers to drain adjoining tracts.[35] Whatever political forces that favored the little guy were now irrelevant.

The majority opinions in the Normanna and Port Acres decisions are masterpieces of opaque reasoning. In both cases, the majority recognized the rule of capture and the mandate of the Hawkins decision to allow each well owner enough production to turn a profit.[36] In the Port Acres opinion, in particular, it is made clear that ample evidence was presented during testimony to establish that, in small tracts in each field, draining from adjoining property was the only way that wells could be operated profitably.[37] Yet, in each case, the court held that Commission orders permitting small tracts to drain from nearby land were violations of property rights and therefore invalid.

It is impossible to divine the intentions of the majority of the justices in thus affirming such opinions. But Justice Smith, in dissenting from each decision, was quite clear-eyed about identifying the motives of his colleagues. The majority of the court had determined, he argued in his Port Acres dissent, that small tracts were an economic embarrassment. Having no faith in the competence of the state legislature or the Railroad Commission to override political pressure and resolve the problem, the court had decided to institute "compulsory pooling by judicial decree."[38] Indeed, this would have to be the effect of the Normanna and Port Acres decisions. Testimony in the latter had established that the drilling and operating of a well would cost $361,674 over the life of a field. If allowed to produce only the gas under a small tract, the owner of that tract would recover little more than $50,000.[39] Obviously, a Commission order forbidding small owners to drain from adjoining property effectively prevented them from applying for well-drilling permits and presented them with a choice between joining pooling agreements or losing all their gas to someone else's well. There is no method of ascertaining whether Justice Smith correctly identified the motives of his colleagues, but the reasoning of the Normanna and Port Acres decisions strongly suggests that his interpretation was correct.

If the majority of the court were trying to force pooling on the producers of Texas, it was probably due to the massive publicity campaign that the majors and larger independents had been waging since the late 1950s. By 1962, it must have been obvious that the Texas industry was in deep trouble and that the only effective way to meet foreign competition was to reduce domestic production costs. As neither the state legislature nor the Railroad Commission had demonstrated a willingness to brave the political opposition of small producers and eliminate favoritism for small tracts, the courts may have felt duty-bound to step in.

DENOUEMENT

At any rate, the Normanna and Port Acres decisions began a peaceful revolution in the Texas industry. It was immediately recognized that they represented a substantial modification of the rule of capture and that, henceforth, the Commission would have to apply new standards of policy making.[40] Truly enough, the Commissioners began a new approach to spacing and allocation. Although the two court decisions pertained to gas fields, their implications for oil fields were inescapable, and the Commission began applying allocation rules based on acreage or acre-feet to both types of field.[41] With these new rules, Rule 37 exceptions became unprofitable, so the small-tract problem was eliminated.

Small independents naturally took all these new rules to court, but by 1964 it was clear that there would be no more discrimination in favor of the little guy.[42] This left small landowners vulnerable to exploitation by the large operators, but in 1965 the legislature passed a compulsory pooling act providing for rules of equity among the participants.[43] Under the pressure of the Normanna and Port Acres decisions, the Commissioners revised not only their specific rules favoring small tracts but their general spacing policy. In 1965 they replaced the 1947 yardstick with a new set of guidelines for allocating production to new fields. Economists agree that this 1965 yardstick gives much greater incentives for wide spacing than did its predecessor.[44]

Since the mid 1960s, therefore, the Texas petroleum industry has not been burdened by favoritism to small landowners and producers in spacing and allocation practices. Not surprisingly, this change was imposed from outside. It was not the democratically accountable Railroad Commission but, rather, the relatively insulated judiciary that had streamlined the state's oil and gas industry.

POLICY-MAKING EPISODE NUMBER FIVE
Slant Wells

In 1920, California officials were surprised to discover that some oil producers had been tapping into the state-owned fields under the Pacific Ocean off the beaches of Southern California. They were surprised because there were no wells out in the water. The oilmen had drilled down on leases near the water's edge, then angled their well bores away from land until they struck the government's pools offshore.[45]

This was the first recorded instance of slant wells. It was significant not only because it illustrated the technological ingenuity in the oil industry but because it involved individuals purloining petroleum. These ingredients of technical inventiveness and disregard for legal ownership would also occur in Texas and would result in a scandal that would embarrass the Railroad Commission and the entire state industry.

The method of curving a well bore is called whipstocking (sometimes whipsticking). A metal sheath, called a whipstock, is inserted at the tip of the drill. This implement causes the rotating drill bit to bite more deeply into the rock on one side of the hole than on the other. As a small boat may be steered in a circular course if you paddle more strongly on one side than on the other, a whipstock will angle the well shaft farther and farther away from the vertical, until a hole that originally pointed straight down begins to travel more nearly parallel to the ground.[46] Aboveground, this deviated well is indistinguishable from a normal well and is capable of producing just as much oil.

There are many legitimate reasons for deviating a well shaft with a whipstock. For instance, it may be much cheaper to drill on the shoreline of a lake, and angle the shaft out to the oil reservoir lying beneath the water, than to construct a platform in the lake directly over the oil. As the California example illustrates, however, there are also illegitimate reasons. There was no legal ambiguity about secretly producing oil from under someone else's property with a slanted well. Early in the history of deviated drilling, the courts made it clear that such an activity was subsurface trespass and therefore forbidden.[47] But, because slanted wells are impossible to detect without special instruments, the fact that they were illegal did not stop them from being drilled.

CONDUCIVE CONDITIONS

For a variety of reasons, East Texas was the ideal place for such a practice to become widely used for the piracy of oil. The crazy-quilt pattern of over twenty thousand wells on hundreds of small and large leases made it impossible for any operator to keep track of the behavior of neighboring operators. Moreover, the geological structure of the field invited slant holes. On the eastern side, the field was bounded by an almost straight line forty-five miles long. The extent and productive depth of the Woodbine sandstone were well known. All an operator had to do was acquire a lease just beyond the eastern boundary, drill a well slanted to the west, and proceed to get rich.

On the west side, theft was somewhat more difficult, but it was still relatively easy. The water drive exerted a constant pressure from west to east, and, as oil was drawn out of wells on the east side, the water advanced slowly toward the east boundary of the field. Gradually, wells on the west side would begin to stop flowing and produce water instead. When this happened, an operator could work over the well, angling it eastward toward the retreating oil, and thus keep producing until the field was exhausted. Most wells need to be worked over at some point, to replace worn pipes, remove sand clogging the bore, or something similar, so the sight of a drilling rig over an old well would cause no suspicion.

Because oil production requires cooperation by drillers, roughnecks, engineers, and others, such practices could not have existed on a large scale without at least toleration from the surrounding community. But slant holing was not only tolerated in East Texas, it was celebrated. By the early 1940s the major oil companies had acquired most of the acreage in the field, especially the best leases on the east side.[48] As the industry later came to realize, the Railroad Commission's spacing and allocation rules allowing wells on small tracts to drain from adjoining leases had had a conditioning effect on the small producers: if they were encouraged to take their neighbors' oil with a straight well, why not with a crooked one?[49] Besides (as some independents in Kilgore will argue even today), the oil under the majors' leases had not been there originally but had been pushed eastward from under other people's leases by the water drive. Why should the majors get the oil just because they had been rich enough to purchase a lease on the east side?

Consequently, draining the oil from under a major company's lease was not considered a crime in Kilgore. As many rural communities encourage moonshiners and despise government revenuers, the people of East Texas encouraged crooked-hole drillers and despised the major companies. Children growing up in Kilgore in the 1950s would commonly hear adults talking about slant wells as they encountered one another in drugstores and barbershops, the way they would discuss the weather or the latest sports championship. There was nothing surreptitious about these conversations; they were normal chitchat on a familiar topic. Kilgore might almost be said to be a community that for nearly twenty years was based on well-organized larceny. Certainly the very substantial participation of important citizens was involved, including two county judges and a county district attorney.[50] And, by the 1950s, the whole operation could not have existed without the participation of at least some of the Kilgore staff of the Railroad Commission.[51]

EVASION

Due to the inexact nature of the art of drilling, it is likely that many of the wells drilled in the early years of East Texas were more or less crooked.[52] But the first well intentionally deviated for purposes of theft was probably put down during World War Two, for a Commission employee later testified that he had heard rumors of such practices in 1946.[53] As time passed and the Railroad Commission attempted with increasing energy to uncover the practice, drillers and producers developed the technology of directional wells to heights of sophistication. By the late 1950s, it was no great feat to bottom a hole as far from the vertical as from the horizontal; that is, a hole drilled four thousand feet deep would also be four thousand feet to the north, east, south, or west. When a full-scale investigation started, one shaft was found to be at an angle of sixty-five degrees, another at fifty-seven degrees.[54]

To complement the technology of directional drilling, operators also refined methods of avoiding detection. For example, they set up dummy wells, connected to slant holes by plastic pipes, which could not be discovered by mine detectors. In Commission records, these fake wells would appear to be successful producers.[55]

They also used marginal or stripper wells to exploit slant holes to the fullest. By state law, wells that produced fewer than 20 barrels a day from the depths relevant to East Texas were exempt from prorationing. By 1960, the Railroad Commission was allowing nonexempt wells to operate only eight days a month. This meant that a well which could produce 21 barrels a day or more was permitted just 168 in a given month, but a well which was capable of extracting only 19 barrels a day, and hence was classed as marginal and was immune to proration, could produce as many as 589. Producers on the watered-out west side of the field, whose wells were down to, say, 3 barrels a day, would have them reworked and deviated out to a point where they would produce, say, 19.75 barrels a day.[56] Just one slanted hole was all that was necessary, for all the other marginal wells could be connected by underground pipes to the one illegal shaft and thereby brought up to just under the production limit for stripper wells.

The amount of oil stolen during the history of East Texas slant holing is not known with precision. Investigators from major companies reported that 60,000 barrels a day were diverted in 1961.[57] Assuming this to be roughly accurate and assuming the pace of operations to have been about the same throughout the 1950s, about 220 million barrels were produced illegally in the decade. As the price of East Texas crude for those years was in the neighborhood of $2.50 a barrel,[58] this would mean that about $550 million was pirated during those ten years. A more common figure for the total value of the stolen oil in East Texas was $100 million,[59] although estimates range from $40 million to $1 billion.[60] Whatever the true totals, they are clearly enormous and probably amount to the greatest outright robbery in history.

THE TECHNOLOGY OF REGULATION, 1947–1960

The slant-well episode was the only major policy-making chapter in the Railroad Commission's history in which there was no dissension among the Commissioners. From the 1940s on, Thompson, Culberson, and Murray were in complete agreement that the intentional drilling of crooked holes was a crime and that it should be prevented if possible and, if not, detected. Decision making on this topic was therefore largely concerned with the technological aspects of regulation. The challenge was not to evaluate and manage conflicting interests but to avoid being hoodwinked by the expertise of the violators.

The Commissioners never intended to try to shut down all the slanted wells in East Texas. They knew that in the early days drilling had been rapid and imprecise. They expected that a large proportion of the older wells, especially those on small tracts (as in the Kilgore townsite), had been accidentally slanted and bottomed under someone else's property. To attempt to detect and eliminate all these wells would be expensive, would embarrass many innocent people, and would create ill will toward the Commission. The Commissioners consequently never issued a general order forbidding slanted wells in East Texas. Instead, they confined their efforts to attempting to prevent new slant holes from being purposely put down into the Woodbine.

This was extremely difficult because the Commission was organizationally unprepared to deal with evasion on the scale occurring in East Texas. It used a staff of auditors and bookkeepers to keep track of production records, but these records were dependent upon the information coming into the office in Austin. In the field, the Commission employed inspectors to oversee its safety and conservation regulations—such as correct well plugging, protection of surface and underground water, and so on—but, as this was generally noncontroversial, it required neither large amounts of money nor a big staff. The legislative appropriation for the Oil and Gas Division of the Railroad Commission in 1960, for example, was just a million and a half dollars.[61] In the 1930s, the Commission had sometimes employed thirty field inspectors in East Texas alone; by the early 1960s, it could afford only three for the entire state, and these had many duties besides searching for crooked holes.[62]

The Commission was short not only of employees but of equipment. When Commissioners began to attempt to prevent deviation in 1947, the agency owned no well-testing instruments. When the Commission received a request for a new well, it could condition the permit on a survey being run by the owner, but it could not run its own tests.

For many years, therefore, the Commission engaged in a losing struggle to acquire adequate information on wells.[63] Beginning in the spring of 1947, Commissioners inserted a straight-hole clause in all Rule 37 permits to drill a well on the east side of the East Texas field—it had not yet occurred to them that there might also be deviated shafts on the west side. This clause required that the driller conduct an inclination survey, witnessed by Commission personnel, and mail the results to Austin.[64] Such a survey involves lowering an instrument called a Totco into the shaft with a cable. A Totco is an aluminum cylinder containing a stable needle on a spring and a spring-timing mechanism that will cause the needle to punch holes in a paper graph at predetermined intervals. The graph is mounted on a counterbalanced rotating wheel that remains oriented to true vertical. Upon being returned to the surface, the disk is inserted in a machine that reads the holes that have been punched by the needle and produces a printout with information on the extent of the well's deviation at each tested depth.[65] Such a survey reveals the shaft's extent of divergence from the vertical but not its direction of travel.

In 1948, the Commission directed its Rule 37 department to consider inserting straight-hole clauses in all permits to drill wells along the eastern *edge* of the east side of the field. Despite these straight-hole clauses, Commissioners continued to hear vague rumors of slanted shafts in East Texas, so in April 1949 they issued statewide Rule 54. This directive required that the Commission be notified of any intentional deviation of a well; in addition, a directional survey had to be filed for any purposefully deviated holes. This relatively expensive test, which became more sophisticated during the 1950s, involves lowering an instrument containing a small camera, a level, and a compass into the well tubing. Every hundred feet, the camera takes a picture of the level and the compass. When the device is brought to the surface and the film developed, the operator has a series of data points which, when plotted, reveal the degree of deviation, the direction of deviation, and, by extension, the bottom-hole location of the well.

The threat of exposure presented by Rule 54 was apparently not enough, however, for a year after its promulgation a district engineer in Kilgore wrote a letter to his superiors in Austin, warning that there was "dire need for additional restrictions on drilling of wells in the East Texas field." The Commissioners responded a month later by issuing a memo to all operators in East Texas, informing them that the Kilgore office of the Commission had been authorized and instructed to conduct an inclination survey, and to require a directional survey, for any reason at any time.

There was soon evidence, however, that all this effort by the Commission was not enough. In the summer of 1950 a Commission engineer, having witnessed a directional survey on a new well the day before, heard a drill running over the new hole late one night. Unable to understand why a completed well would still require a drill, he contacted his superiors in Austin. They ordered a resurvey of the well.

The investigation revealed that the drilling and survey companies had been cooperating to deceive the Commission. Once a straight hole had been drilled and tested in front of witnesses, the drill was set up again, and a second, deviated shaft was pushed out to the oil-rich portion of the Woodbine. In Commission records, this well would have appeared as certifiably straight.

The Commissioners both severed this well's pipeline connection and began legal action against the drilling company.[66] Then, in January 1951, they issued a special order mandating a directional survey for *all* wells to be drilled on the eastern edge of the East Texas field. Moreover, they began instituting a series of informal requirements, tightening Rule 54.

First, they stipulated that the directional survey had to be run no more than forty-eight hours before the well actually began to produce. As a new hole could not be drilled in only two days, it was expected that this requirement would guarantee that the survey reports were from the correct shafts. Second, they purchased inclination survey equipment for the Commission and began to run their own surveys. And, third, they rotated the engineers who ran these inclination surveys around the state, on the theory that constant movement would prevent them from becoming familiar enough with the people in East Texas to become corruptible.

By the late 1950s, a newly drilled well on the east side of the East Texas field had to survive a complicated series of procedural checks in order to be deemed acceptable by the Commission. Two inclination surveys were run, one by the Commission and one by the drilling contractor. A directional survey was performed by a private company on an approved list. The records for these surveys, submitted to the Austin office under oath via registered mail, were checked against one another by the Commission's staff.

Still, reports coming into Austin from this elaborate system of oversight during the decade all showed relatively straight holes. In evaluating this information, Commissioners were faced with a choice. They could assume that there was a tightly organized conspiracy at work, involving producers, drillers, surveying companies, and Commission personnel, or they could discount the possibility of further deviated holes. Having confidence in their field inspectors, they chose to disregard the persistent rumors of slant wells in East Texas.

SUSPICION: 1961

As a new decade dawned, Ernest Thompson and Olin Culberson became gravely ill. Culberson's death in 1961 brought an inexperienced Ben Ramsey to the Commission. Ramsey opposed slant wells but had not yet acquired the expertise to understand the situation, and Thompson was frequently absent from Commission hearings, so the entire problem of dealing with slant holes in the 1960s fell to William Murray.[67]

This was no small burden, for in April 1961 a Shell crew working on an old well on the west side of the East Texas field noticed that fresh drilling mud was pulsating out of their casinghead. Drill bits are lubricated with mud. It was obvious that such a thing could happen only if the drill bit from another well had pierced the Shell casing underground. The mud must have entered the other well at the surface, traveled down the new well's shaft, transferred to the old well's casing through the break in the pipe, and been forced to the surface. The only drilling rig in the area was over half a mile away, on someone else's property. A Commission engineer who happened to drive by noticed the incident and reported it to Commissioner Murray in Austin.[68]

Although the full force of the scandal was not to break for another year, this episode effectively ensured the demise of slant holing. The major companies, led by Humble, began a private investigation.[69] Several newspaper reporters began nosing around the edges of the conspiracy. And William Murray came to two realizations. First, the technique of slant-hole drilling had advanced far more than engineers like himself had been aware. Up to that time, petroleum engineering students had been taught that, as a rule of thumb, twenty-five degrees was a maximum deviation for a slanted well. But this assumed a specific target for the drill. It was not realized that, if a driller just pointed the shaft in a general direction, the well could deviate far more than the theoretical maximum. Second, Murray realized that there were deviated wells on the west as well as on the east side of the field.

Accompanying these realizations was the suspicion—which was not confirmed until much later, after resurveys of wells that had been tested by Commission personnel—that at least some of the Railroad Commission field inspectors in East Texas might be corrupt. And, consequently, the survey data in the files of the Commission, showing straight well after straight well, could possibly be fraudulent.[70]

The regulatory problem thus became more difficult. The Commission still had only a very limited budget; now the Commissioners did not know how much confidence they could place in their staff. From Austin, Murray had to attempt to discover whether the Commission survey reports had been faked and, if so, how, as well as how the slanted wells might be identified. To make matters worse, the Normanna decision had been handed down in March, and the Commission had to devise a new spacing and allocation policy. Throughout the summer and fall of 1961, Murray and the new employees he hired because they had never been to East Texas struggled with these problems. Meanwhile, the private investigations proceeded independently.

SCANDAL: 1962

In January of 1962, Commissioner Murray had an inspiration. It seemed likely that producers who would drill slant wells would try to get the maximum out of them. They would therefore want it to be classified as a marginal well and, as a result, be exempt from prorationing. Further, they would want it to produce up to the twenty-barrel limit for a stripper. But a real marginal well would decline steadily in production from year to year. Consequently, the slanted fake strippers should be easily spotted by their consistent production of just under twenty barrels a day.[71]

There were 3,895 marginal wells in East Texas, each with production affidavits in Austin.[72] Murray hired a University of Texas engineering student part-time to examine the records of each of these wells and note every one whose production remained at almost twenty barrels a day from year to year. He took this list of several hundred suspicious wells to the Texas attorney general's office and requested an investigation. There he was told that the attorney general could not authorize such a fishing expedition and that he should come back when he had evidence rather than inferences.[73]

There the matter might have rested, but on April 11, 1962, a reporter named Jim Drummond brought the story to public attention with an article in the *Oil Daily*.[74] Slant holes suddenly became a hot news topic, and more stories began appearing in other papers.[75] With this rash of publicity, the search for deviated wells was no longer a fishing expedition, and the attorney general's office began a vigorous investigation. In the next two months, with the Railroad Commission, the attorney general, the state legislature, and, eventually, the United States Department of the Interior all pressing investigations, the scandal broke fully, and sensational revelations dominated the Texas industry's attention for the rest of the summer.[76]

It is not necessary to follow the course of these investigations in detail, but some aspects of the process are important. The guilty operators of the East Texas area naturally attempted to evade punishment—with some success, for local juries were often sympathetic.[77] More relevantly, however, they actually launched a political offensive against the major companies and the Railroad Commission. On June 6, 1962, about eighty operators met in Kilgore and formed the East Texas Producers Group, whose purpose was, in essence, to discredit the forces seeking to expose the slant wells. Their strategy was to suggest that the crooked-well investigations were a ploy by the majors to squeeze independents. They charged that the Commission was a tool of the majors and called upon the United States attorney general to investigate the majors for antitrust violations. Nothing much came of this campaign, but the fact that it was launched at all illustrates the bitter divisions that continued to exist within the Texas industry.[78]

As the truth came out during the summer, William Murray learned how the slant holers had evaded the Commission's detection system. The supposedly trustworthy crews from private companies who ran directional surveys had been running preexposed film, showing a straight hole, through their equipment. The film readings in the Railroad Commission files were thus for the wrong wells.

The fraud perpetrated by the Commission staff had been more subtle. The crews who ran the inclination surveys for the Commission had been honest; it was some of the Kilgore field personnel who were part of the conspiracy. Running a survey is a dirty job, and the crew members would get earth and grease on their hands during the operation. The field inspector, his hands clean, waited in the truck while they performed their job. After taking the well readings, the crew would give the survey disk containing the raw information to the field inspector while they wiped their hands. The inspector would simply substitute a survey disk taken from another well for the one the crew had just made. Their hands cleaned, the people in the crew would sign the Commission forms attesting to the disk's reliability, then innocently mail the incorrect form and the phony disk to Austin. Again, the records in the Railroad Commission files were for the wrong wells.

Although it is evident that there must have been substantial bribes involved in these operations, the courts were able to come up with little information about their extent.[79]

THE PROBLEMS OF REGULATION

The great slant-well scandal could occur because the Railroad Commission was unable to effectively supervise the means of producing oil in Texas. Partly, this was the result of the agency's inadequate budget. Partly, it reflected the difficulty of ensuring that public employees would remain honest when confronted with what must have been spectacular bribes. And, partly, it arose from the disadvantages at which a regulatory commission must find itself when confronted with a resourceful population employing technological innovations. Because it could not find adequate equipment or personnel, because the people it did employ were confronted with overawing temptation, and because it could not keep up with the evolution of technology, the Commission failed to halt the spread of slant wells for nearly two decades.

This inadequacy was not due to any reluctance among the Commissioners to deal with the problem. Nor was it caused by political or other disagreements about the proper means to approach it. Rather, the Commission's failure is simply another example of the common inability of regulatory agencies to adequately control the industry for which they are responsible.[80]

The Railroad Commission was able to establish control of the industry in the 1930s because the industry wanted regulation. The major companies always wanted it, and the independents came to want it after the general thrust of the Commission's spacing and allocation policy became clear. But slant-well producers did not want to be controlled. They were able to avoid detection until a stroke of misfortune—the Shell incident in 1961—set off a chain of events that finally exposed them. But the general forces that led to easy evasion were not ended by the investigation. There are even today rumors in East Texas about slant wells still pumping happily almost two decades after they were supposed to have been eliminated. Whether or not these rumors are true, the general circumstances surrounding the scandal suggest pessimistic conclusions about the power of a government agency to regulate a dynamic industry when that industry resists regulation.

THE SCANDAL'S PRICE

Considering the sound and fury that accompanied the revelations about East Texas drilling practices, the outcome was relatively insignificant. Three hundred eighty deviated wells were eventually discovered and shut down in East Texas, most of them just off the eastern edge of the field.[81] Twenty-three crooked holes were found in the nearby Hawkins field. Slightly over one million dollars was collected by the state in civil penalty suits.[82] Humble, Texaco, Conoco, and several other major companies filed suits, but the courts ruled that they could recover only two years of damages, which severely limited the amount of money they could expect. Most of these suits were settled out of court, for undisclosed amounts.[83] Although criminal charges were filed against more than one hundred sixty people, only one was convicted, and that conviction was overturned.[84] Some of the principal thieves still live comfortably in Kilgore, as respected members of the community.

The greatest price was paid not by the slant holers but by the Railroad Commission. William Murray made enemies during the early 1960s, both because he pursued the truth in East Texas and because he espoused natural gas prorationing (more about this in chapter 5). It is commonly believed in the Texas industry that in 1962 certain operators in East Texas set out to get Murray before his investigation could hurt them. Although, by the summer of 1962, no individual could have stopped the unfolding revelations, the operators succeeded in getting Murray.

In the early 1950s, Murray had been affiliated with a drilling firm operating in West and North Texas. In 1953, the firm had acquired some producing leases in the Spraberry field. Despite the fact that the Railroad Commission did not regulate drilling companies directly, and despite the fact that Murray had sold his interest in the drilling firm shortly after it had become a producer, such a relationship clearly raised the specter of conflict of interest. Moreover, in the late 1950s, Murray had been involved in a land deal pertaining to some oil leases that had, on paper, made him a profit of over a million dollars. This also did not involve Commission regulations directly, but, again, it included relationships of a questionable character for a public servant.

The muckraking *Texas Observer* had published accounts of these transactions when they occurred, but the state's establishment press had ignored them. In the spring of 1963, however, the *Dallas News* published a series of articles detailing Murray's business dealings and equating them with the slant-hole scandal.[85] Although all his questionable activities had ceased years before, and although an Austin grand jury refused to indict him, weeks of bad publicity stained both Murray and the Railroad Commission.[86] He resigned in April. This was perhaps the most important lasting result of the slant-well scandal.

The Murray resignation illustrates a critical problem for any regulatory agency that deals with a technologically developed industry like petroleum. By consensus, Murray is judged the most technically competent individual ever to serve on the Commission. In the early 1950s, the yearly salary of a Railroad Commissioner was under seven thousand dollars.[87] This was not such a problem for the other two Commissioners, for they had outside business interests: Ernest Thompson owned a hotel in Amarillo, and Olin Culberson owned a general store in Edna. But Murray, professionally trained as a petroleum engineer, could hardly be expected to seek outside income except in the oil and gas business. The very training that made him a capable Commissioner, therefore, condemned him either to live a life of relative penury or to skirt about the edges of conflict of interest. Murray's ordeal in 1963 makes it unlikely that other participants in the industry will come to the Railroad Commission. Certainly, they will never arrive through appointment and, probably, never through election. This may reassure members of the public about the integrity of their servants, but it does nothing to enhance the technical qualifications of the people who regulate the industry. In the 1970s, many members of the industry complained privately that the quality of the agency had deteriorated, because none of the three Commissioners had technical training. As one field inspector put it in 1978, "Every time we go in to see them, we have to educate them all over again." The charge of conflict of interest aimed at Murray was dramatic, but it missed the real issue: how to discover a means to get able individuals on the Commission, without their having ties to the industry they regulate.

5. Welcome to the Energy Crisis, 1965–1980

We do not know what is happening to us, and that is precisely the thing that is happening to us—the fact of not knowing what is happening to us.—José Ortega y Gasset, *Man and Crisis*

Ernest Thompson resigned from the Railroad Commission in January 1965 and died the next summer. Governor John Connally appointed Texas House Speaker Byron Tunnell to his seat. With Culberson's death in 1961 and Murray's resignation in 1963, the Commission had experienced a complete personnel changeover in four years. This was in marked contrast to its stability in the previous decade. In retrospect, the Commission's rapid turnover in the early 1960s can be seen as a symbolic curtain raiser to the energy turmoil that would strike the agency, the nation, and the world in the coming years. Thompson's resignation is as good a point as any to mark the end of the heroic era in the Texas petroleum industry and the beginning of its transition into a historical epoch marked by disruption and perplexity.

Thompson's resignation did not deprive the Texas industry of its spokesman to the outside world. Jim Langdon, appointed by Connally to replace Murray in 1963, had assumed the mantle of the Railroad Commission's public representative from the ailing Thompson. Langdon was just as eloquent, energetic, and self-confident as Thompson, although he lacked the colonel's colorful charm. It was Langdon who took it upon himself to reassure small producers and royalty owners that the Commission was still on their side and that it was determined to enforce equity in the state's petroleum fields, after the Normanna and Port Acres decisions.[1] And it was Langdon who traveled to an increasingly hostile Washington, D.C., to present the position of the Texas industry in the energy wrangles of the late 1960s and early 1970s.

By the time Langdon left the Commission at the end of 1977, history had heightened the role of Commissioners as representatives of the Texas industry. The energy crisis had focused so much attention on Texas producers that all the men who served in the late

1970s—Mack Wallace, Jon Newton, John Poerner, and Jim Nugent—were almost constantly making speeches or testifying before Congress. The Commission, which in previous years had been well known in the world of petroleum and almost unknown outside of it, was becoming the object of increasing attention from consuming interests in the northern industrial states, as well as inside Texas. The old politics of energy was a politics of surplus, and the disagreements it spawned, while often intense, were relatively uninteresting to the public at large. But the new politics of energy is a politics of scarcity, and that fact has transformed the debate over national energy policy into front-page news.

Since March 1972, the Railroad Commission has not prorated oil to market demand; with insignificant exceptions, all allowables are now set at 100 percent. The Commission's decisions are consequently no longer crucial to United States oil supply. Texas still furnishes, however, about a third of the nation's natural gas, and the battles over the allocation of that resource have been as heated as any over oil in the past and of far wider scope.

Comfortable, informal, personal methods of administration have crumbled as Commissioners have found themselves in a fishbowl, constantly the subject of often hostile publicity. In the days of surplus, no one really minded informal policy making, but the onset of mass protest, the frequent television coverage, and an intrusive federal government have made it necessary to tone down Texas' image of being dominated by Stetson-wearing wheeler-dealers in smoke-filled rooms. In the mid 1970s, the state legislature passed the Administrative Procedures Act and the Open Meetings Act, which forbade Commissioners to discuss any policy privately with anyone, even among themselves. All Commission deliberations now have to be carried on in public. Jim Langdon, a politician of the old school, could not adjust to such methods and, partly for this reason, resigned from the Commission in 1977. Commissioners who have served since then dislike the public deliberation laws, but they endure them.[2]

With the increase in visibility has come an expansion of responsibility. The rising prices of oil and gas have sparked interest and investment in other energy sources. Their exploitation, like that of petroleum, poses environmental problems and must be regulated. In the 1970s, the state legislature gave the Railroad Commission responsibility for overseeing lignite coal mining and the extraction of energy from geothermal resources. This has involved the Commission in new varieties of political choice—but that is beyond the scope of this book.

As the 1980s begin, Commissioners are in the final stages of dealing with two problems that were thrust upon them during the first shock of the energy crisis. Both concern natural gas. The first involves their effort to prorate gas production following the surplus that appeared in the intrastate market in 1975. The second entails their attempt to cope with the inability of the state's largest utility to meet its obligations to consumers.

POLICY-MAKING EPISODE NUMBER SIX
Gas Prorationing

The problems of prorating oil production are abstruse, complex, and divisive, but they are nothing compared to the difficulties of controlling natural gas. First, about a third of the gas supply is nearly beyond influence, for it is casinghead gas that is produced automatically with oil. Second, natural gas is extraordinarily difficult to transport and store. Unlike oil, it cannot be stored in open containers, and, also unlike oil, its transportation in anything but a pipeline is inordinately expensive. Third, the demand for gas fluctuates markedly with the season, being high in winter and summer and low in spring and fall.[3] Fourth, the price of gas was for a long time extraordinarily low, compared to the price of other fuels. Even as late as 1960, Texas gas sold for just over ten cents a thousand cubic feet, which made it less than one-fourth as costly as oil for the same heating capacity.[4]

All these circumstances combined to force the gas industry into an economic structure somewhat different from the one experienced by the oil industry. Natural gas producers have to sell their product to a pipeline company; unlike their oil-producing brethren, they cannot truck it to market. Consequently, there are normally very few, or even just one, buyers of gas in a field, whereas there is a potentially unlimited number of buyers in an oil field. Gas producers are therefore much more vulnerable to pipeline company buying policies than are oil producers.

With the price of gas so low, the market seasonably changeable, and transportation dependent upon the pipeline, producers needed assurances that they would have a steady, guaranteed buyer for their output, or they would be unable to get bank financing to drill. In contrast to oil, where there was a fluctuating posted price adhered to by the many buyers in a field, therefore, gas was produced under long-term contract to a single pipeline company. Even where there was more than one pipeline to a field, there was little short-term competition for gas; once contracts were signed, the production and transportation system of any given field was set for years.[5] This introduced rigidities into the structure of gas supply and prevented the industry from adjusting rapidly to market forces.

The gas industry, however, was burdened by the same problem of oversupply as was the oil industry. Gas was also subject to the rule of capture, and competing operators would, if left to themselves, drain each others' land to keep from being drained. In the "normal" situation of surplus that prevailed before 1973, gas producers, like oil producers, wished to be prorated to preserve equity among themselves. But, because of an unfortunate set of legal and economic circumstances in the early 1930s, the pattern of gas prorationing evolved in a much different direction from the system of oil production.

THE OLD SYSTEM: ROUTINIZED CHAOS

While it was attempting to bring order to the oil situation in the early 1930s, the Railroad Commission was also trying to control the gas industry. Unlike the determined self-confidence that Ernest Thompson showed in dealing with oil, however, he treated gas uncertainly and timidly. In a series of decisions after 1935, state and federal courts overturned Commission proration orders and issued injunctions against further attempts to control gas.[6] During the oil proration struggle, the Commission had lost many similar court battles but had continuously maneuvered until it had secured judicial acquiescence to its authority. In contrast, the Commission accepted defeat in the 1930s and allowed the gas situation to drift. While the oil prorationing policy underwent an evolution that was acceptable to both the Commissioners and the industry, gas prorationing remained merely a potential.

In 1944, however, two operators in the Bammel gas field near Houston could not agree on mutual allocation of the field's resources, and one sued to force the Commission to step in. In the case of Corzelius versus Harrell, decided the following year, the Texas Supreme Court ordered the Commission to allocate production in

Bammel.[7] Shortly thereafter, a federal court lifted the injunction forbidding the Commission to prorate gas. Commissioners were once more authorized to regulate this portion of the petroleum industry.

But, by 1945, a private system of gas prorationing had grown up parallel to the public system of oil prorationing. For three reasons, this system rested on the setting of rules for individual fields rather than on a statewide, across-the-board scheme, as with oil.

First, there was very little interconnection of gas pipelines. Oil companies, enmeshed in a crisscrossing, interlocking system of transportation, could shuttle their product around in a complicated system of trade-offs. This was impossible with gas.

Second, the price of gas varied much more widely than the price of oil. Because gas contracts were made between pipelines and producers for long periods, the price did not fluctuate and adjust to an overall equilibrium, as it did with oil. The posted price of oil was very similar in different areas and fluctuated with the market demand each month. There was no posted price for gas, so the product carried by different pipeline companies was not interchangeable.

Third, because of the paralysis induced in the Railroad Commission by the court decisions of the 1930s, no effort had been made to overcome these problems at their inception. By the time the Commission officially assumed statewide authority over gas in 1945, traditions of operation were already established in the gas industry.

For these reasons, the Commissioners of the late 1940s set about to work *within* the prevailing private system of gas prorationing. They tried to ensure that pipeline companies treated producers in connected fields fairly by issuing ratable-take orders forcing purchasers to buy proportionately from all producers, but they made no attempt to impose blanket regulations on the industry. Gas prorationing therefore became a matter to be constructed afresh for each field, depending on the physical circumstances of that field, the established economic relationships, the expressed wishes of the operators, and the Commissioners' own notions of fairness.

The result was chaos. The Commission became involved in setting production rules for a field only if asked to by its operators. In many fields, especially those with just one or a few large operators, the Commission was not asked to intervene, and the producers devised their own rules. If the producers asked for Commission participation, they could suggest a variety of allocation formulas. In some fields production was allocated on acreage, in some on bottom-hole pressure times acreage, in some on acre-feet times bottom-hole pressure, and so on.[8] If there were small tracts in the field, however, the Commission decreed allocation based one-third on the well and two-thirds on acreage, thereby allowing small tracts to drain their neighbors.

Thus for the allocation of production to wells. But this did not regulate the pattern of purchases among the pipelines connected to the field. If a field contained only one pipeline, the Commission did not attempt to fix its take. But, if two or more lines serviced a field, a different demand from each could cause drainage across property lines. In that case, the Commission would set an allowable for the field that fell midway between the needs of the pipelines.

For example, suppose pipeline A wanted to take 70 percent of the theoretical maximum from, say, half the wells in a given field, and pipeline B wished to take 30 percent of the theoretical maximum from the other half of the wells. In that case, operators and royalty owners connected to pipeline B were in danger of having their property drained by wells connected to pipeline A. Under such hypothetical circumstances, the Commission would be apt to set the field allowable at 50 percent.

Often, under conditions like these, the two pipelines connected to the field would adjust their production to mutual advantage—that is, A would buy gas from B. Sometimes, however, technical problems or federal regulations (to be explained shortly) would interfere with such an easy solution, and no satisfactory arrangement could be achieved.

This system had far different consequences from the one created to control oil production. In oil, allowables were set for the entire state. There was therefore a constant activity of balancing *between* fields, as purchasers with too high an allowable sold oil to those with too low an allowable. In gas, the adjustments took place only *within* fields. Because there was no statewide, across-the-board proration system for gas, the Commission could not force gas purchasers to swap allowables among fields and areas of the state. This made the gas industry fundamentally different from the oil industry in two ways.

First, wildcatters who discovered gas in small, remote, or otherwise undesirable fields were sometimes unable to get a pipeline connection. If no pipelines wanted to extend to them, the Commission could not come to their aid by refusing to raise allowables. Because major companies had their own pipelines, this situation was not a problem for them. Independents, however, were often highly vulnerable under such a system.

Second, the setting of production allowables on a field-by-field basis resulted in an almost infinite elaboration of rules. Ratable-take orders, special allowables, seasonal adjustments, and balancing allowables proliferated.[9] In the absence of a statewide proration system to allocate production to all fields and all wells according to some comprehensible formula, the exceptions and special rules multiplied beyond the point of understanding even by those closest to the process.

To these complexities was added, in 1954, the problem of inter- and intrastate gas. In the Phillips case, the U.S. Supreme Court in that year ruled that the Federal Power Commission had the authority and the duty to regulate the wellhead price of natural gas destined for interstate commerce.[10] All the other circumstances inherent to the petroleum industry had made the creation of a rational system of gas regulation in Texas unlikely. The Phillips decision made it impossible.

One of the facets of the long-term contracts between gas producers and pipeline companies was a common provision known as a take-or-pay clause. Designed to ensure a steady market for producers, so that they could acquire bank financing, these contracts provided that a pipeline would take a certain volume of gas from a producer or pay that producer as though it had taken the gas.

If a state conservation commission ordered an interstate pipeline to cut its purchases to preserve ratable take, thus forcing that company to pay a penalty to its producers, the pipeline would have to raise the price of its gas to recover its losses. But this meant that a state agency was (albeit indirectly) regulating the price of interstate gas.

The Federal Power Commission objected to this practice, and, in 1963, in the Kansas case, the U.S. Supreme Court forbade state commissions to thus interfere with interstate commerce.[11] The court did not prohibit state agencies from setting *production* levels for wells connected to interstate lines, but the state could not regulate the amount that those lines could *purchase*. The Railroad Commission was thereafter free to attempt to achieve equity between wells in fields served by both types of pipeline by juggling allowables but not by directing them to take ratably. An essential piece of regulatory power was thereby denied it, and the problem of gas prorationing grew ever more intractable.

But that was not the end. Under the authority granted it in the Phillips decision, the FPC decreed that, if *any* gas from a given well were sold to an interstate system *at any time,* it was forever interstate gas and thus was subject to FPC price-fixing. Since the FPC persisted in setting prices below the market level in Texas, intrastate lines avoided selling gas to interstate lines, for fear of falling under FPC authority.

This meant that, in a case like the hypothetical one already mentioned, in which pipeline A wanted more than its allowable and pipeline B wanted less, cooperation between the two was prevented if one were an inter- and the other an intrastate line. If B were the intrastate line, it was not about to sell gas to A, because that would bring it under FPC controls. The two companies would continue to take the same gas from the same field, each aware of the benefits of cooperation but unable to do business because of the wall of separation between the state and federal markets.

All through the 1950s and early 1960s, the Railroad Commission attempted to bring some order to this tangle of regulation. In this effort it was encouraged by the independent producers, who felt victimized by the system. They especially disliked the fact that there was no statewide allocation system, as there was for oil. The Commission could compel the major companies to take oil ratably from all fields in the state by simply refusing to raise allowables on a field-by-field basis. With the oil production system based on statewide production, purchasers were forced to connect to new fields if they wanted more oil. But, because there was no similar system for gas, pipelines could buy from conveniently situated fields and ignore others. This vulnerability to pipeline prorationing made the independents jittery, and they looked to the Commission to protect their interests.

In the late 1950s and the early 1960s, therefore, the independents put on a campaign to get the state legislature and/or the Railroad Commission to overhaul Texas' gas regulation system. Naturally, pipeline company representatives opposed a change to across-the-board prorationing, so gas producers and transporters made much unfriendly noise in industry convention halls and in the pages of industry journals for nearly a decade.[12]

Olin Culberson, Ernest Thompson, and William Murray were, as usual, sympathetic to the independents on this issue, but the problem's complexity kept them from devising a comprehensive solution. Throughout the 1950s, they contented themselves with refining ratable-take orders for individual fields.[13] In 1963, however,

Murray formulated an across-the-board proration plan which he was going to ask his fellow Commissioners to endorse.[14] If such a scheme had been adopted in 1963, it is possible that many of the dislocations of the 1970s could have been avoided. But the bad publicity about Murray's business ventures hit just after the plan's announcement, and, when he resigned, the statewide gas proration system went down with him. After 1963, gas proration in Texas was allowed to drift, to be taken up again by a less knowledgeable and more harried Commission in the mid 1970s.

SURPLUS AMID SCARCITY

When the natural gas shortage became acute in 1973, the price of that commodity began to rise rapidly on the intrastate market. The average price per thousand cubic feet, which had been 14.4 cents in 1970, jumped to 20.4 cents in 1973 and to 51.9 cents in 1975.[15] Of course, gas committed under previously existing contracts still sold at preshortage prices. Suddenly, gas exploration was potentially profitable, and wildcatters rushed to drill. There had been only 763 successful gas well completions in 1968, but that number climbed to 1,475 in 1973 and 2,135 in 1975.[16] Because the Federal Power Commission continued to keep the price of interstate gas relatively low, producers all attempted to sell their newly discovered gas on the intrastate market. Because of the apparent gas shortage, Texas gas utilities signed hundreds of contracts and asked for more.

But, in 1975, a peculiar thing happened. Because higher prices had forced conservation on the public, because the Railroad Commission had earlier ordered industry to cut back on the use of gas as a boiler fuel, and because increased exploration had leveled off the previously rapid decline in reserves, the intrastate pipelines discovered that they had very much more gas on hand than they required immediately. While the rest of the country endured a gas famine, Texas experienced a surplus. By 1977, this unused surplus had reached 606 billion cubic feet and was headed upward.[17]

If the surplus had been in oil, it would have forced the price down. But the excess gas supply was already committed by contracts to intrastate lines. Market forces could not pressure the price downward, the market having been preempted by long-term contracts.

As the surplus developed, and as pipelines discovered that they could not sell all their gas, they began to cut back on their purchases. But this involved all the inequities of pipeline proration:

they took gas from those producers with whom they had long-standing cheap contracts and ignored the operators with whom they had more recently signed more expensive contracts; they took only from producers with whom they had take-or-pay obligations; they took from some fields and not others, and so on.[18] Producers could not sell their remaining gas elsewhere, because it was contractually committed to the company that was not taking it and because, without the pipeline, transportation was impossible. Their wells shut in, producers beseeched the Railroad Commission to save them from ruin.

Despite the national shortage, therefore, Commissioners were forced to attempt to distribute the purchases of intrastate gas ratably, which inevitably involved suppressing production. In February 1977, the Commission ordered the prorationing of all gas fields in Texas.[19] The impact in Washington of the news that Texans were curtailing gas production can be imagined. In February 1978, Senator Henry Jackson, appearing on "Face the Nation," accused the Railroad Commission of price-fixing and suggested federal confiscation of Texas' gas.[20] In March, a Department of Energy official threatened dire consequences for Texans.[21] Even within the state, consumer representatives were unhappy.[22]

It is, of course, difficult to explain to nonspecialists, especially those enduring a northeastern winter, that intrastate gas prorationing is not a price-fixing scheme but is merely a series of technical adjustments to protect producers. Commissioner Mack Wallace attempted such an explanation, with indifferent success.[23] And, as the battle over President Carter's energy program testified, representatives of the consuming states are in no mood to trust Texas' good intentions when it comes to petroleum policy.

OBSTACLES

Despite the external criticism, however, the major problems with the gas prorationing effort were internal. Part of these were technical. The Commission ordered its staff to get every field prorated by the end of 1977. But as each field required a separate hearing, and as there were hundreds of unprorated fields, the process took much longer than anticipated. By the end of 1979, all fields were still not included in the system.

Other problems were structural and, therefore, more serious. The Commission of the 1970s inherited the jerry-built, inefficient, field-by-field gas proration system of the 1950s. Instead of throwing out this whole hodgepodge and beginning over with statewide allowables, the Commissioners devised an even more complicated arrangement of prorating by fields. Each producer in a field was to file a forecast of the estimated demand for the coming month, and each purchaser (the pipeline) was to likewise file a nomination for the expected take. Using a computer, the Commission staff would then calculate an allowable for the field based upon an adjustment of the two estimates.[24]

In theory, this procedure would balance production to demand, and everyone would be happy. In practice, the system has been a fiasco, because both producer and pipeline estimates have been grossly inflated. Clearly, asking producers to estimate the demand for their own gas gives them an incentive to be wildly optimistic. On the other hand, it is not optimism but caution that induces pipelines to overnominate. Fear of being caught short on their gas supply causes them to nominate up to the very limit of the range of their actual potential demand. When these two forecasts are combined in the Railroad Commission computer, they produce allowables that have been, in practice, at least 50 percent above the eventual market demand for the field. As computer programmers say: garbage in, garbage out.

The Commissioners of the 1950s understood the tendency of both producers and buyers to overestimate demand for oil and, therefore, took precautions against it. No producers' forecasts were solicited, and purchasers' nominations were of secondary importance, in setting allowables, to records of crude-oil inventories. But the members of the latest Commission have not learned to discount industry forecasts, and consequently they have failed to match supply to demand.

When the Commission sets gas allowables too high, the purpose of government prorationing is defeated. Pipelines, faced with more than they can handle, initiate their own system of buying, with all the attendant hardships on producers that have been discussed in the present chapter. In its response to the intrastate gas surplus of the late 1970s, therefore, the Commission has brought upon itself all of the disadvantages, in terms of outside criticism, and none of the advantages, in terms of supply stability, of a system of government control of production.

As the 1980s begin, the anarchy of gas prorationing could be solved by breaking with the patterns of the past and moving to statewide prorationing, similar to the system that worked so smoothly for oil. Until Congress passed the Natural Gas Policy Act in March 1978, such a rational plan was prevented by the differences between intra- and interstate gas. With the removal of barriers to interstate sale by intrastate pipelines, and with the increase in the permitted price, however, the way has been cleared for a comprehensive approach to the problem of gas production in Texas.

The major obstacle now exists not without but within the Commission, or, more accurately, within the Commissioners. Some producers argue that to institute across-the-board prorationing would interfere with established take-or-pay contracts. They feel that if the Commission, in adjusting statewide production to statewide demand, ordered a pipeline's take held below a level for which it had contracted, it would be forced to pay a stipulated penalty. The Railroad Commission, they maintain, cannot legally interfere with established contracts.

That members of the petroleum fraternity should thus argue is understandable. That Commissioners should agree with them is harder to accept. In the glory days of oil prorationing, the Commissioners adjusted allowables to demand on a statewide basis and forced the industry to conform to their dictates irrespective of contractual obligations. After 1934, the Commission was consistently upheld in this authority by the courts. After 1945, it was upheld in its right to prorate gas. This makes good legal sense: a trucking firm, to take an analogous example, could not claim that it had contracted to deliver goods at seventy miles an hour and therefore was exempt from state speed laws. Just so, it is perplexing to hear industry representatives argue that, because pipelines have contracted to buy a fixed quantity of gas, they are therefore exempt from state prorationing. Yet this is what they claim, and Commissioners have been agreeing with them.

The explanation for this singular position seems to lie in the differing ideological makeup of Commissioners of the late 1970s, as opposed to that of their predecessors. The men who regulated the Texas industry in its decades of dominance were very friendly to business, but they never felt the compulsion to endorse the concept that business is most prosperous if left alone. Indeed, the crisis of the 1930s had convinced them that the oil industry needed a firm guiding hand. Nor did they seem to believe in the sanctity of private property. The prosperity of Texas was their oft stated goal, and they were not timid about using their authority to force private property owners to bring prosperity to the state.

But the present Commission seems to contain men who have

been seduced by the antiregulation rhetoric of recent years. They are reluctant to exercise their own indisputable authority. As one Commissioner stated in answer to a gas producer's question about prorationing at an industry trade association's state convention in June 1978:

> There's a matter of us regulating, but there's also a matter of us interfering with private business and private enterprise and not going so far as to override individual contracts between people, producers, and pipelines.[25]

It is inconceivable that a Commissioner of twenty or forty years ago would have thus publicly expressed doubt about the Commission's ability to regulate the petroleum industry for the public good.

SUSPENSION

The problem of the intrastate gas surplus was temporarily suspended in 1978 by the passage of the Natural Gas Policy Act. This federal law authorized the sale of intrastate gas to the interstate pipelines and protected them from falling permanently under Department of Energy authority by making such sales. With this new freedom, intrastate gas began moving in interstate commerce, and, by the late summer of 1979, the Texas gas surplus had been absorbed.[26] With the problem gone, any pressure to overhaul the state's gas proration system also eased. But this respite from the gas proration predicament would be short-lived. A surplus gas bubble reappeared in the late spring of 1980 and it began to look as though even interstate sales would not solve the problem of too much gas in selected fields.

Whether or not the 1980 gas surplus will be distributed by short-term expedients, the long-term problem—lack of statewide prorationing—remains. Drilling and technological innovation continue apace, and higher prices may curb consumption, so it is possible, even likely, that the nation will experience frequent spot surpluses in the future. When that happens, the Railroad Commission will again be faced with trying to adjust an antiquated system of gas prorationing to modern conditions. Its handling of the problem in the 1970s does not suggest an optimistic outlook for the 1980s.

POLICY-MAKING EPISODE NUMBER SEVEN
The Lo-Vaca Problem

In many respects, the Lo-Vaca quagmire of the 1970s was similar to the oil proration crisis of the 1930s. In both cases, a relatively staid and comfortable agency suddenly became responsible for a new and seemingly insoluble problem. In both cases, political pressure of an intense and novel sort was focused on the Commissioners. In both cases, the Commissioners who were first faced with the problem were not up to dealing with it. And, in both cases, the eventual solution involved the Commission in greater and more complex regulatory powers.

The Lo-Vaca problem caused the Commissioners to concentrate more on the regulation of gas utilities than on the production of petroleum. For the generation preceding 1973, the most important department of the Commission was its Oil and Gas Division. Commissioners, accustomed to arbitrating disputes within the petroleum industry, were forced to deal with outside troubles in only a peripheral way. With the announced failure of the Lo-Vaca Gathering Company to meet its obligations to supply natural gas to a third of the population of Texas, however, the Gas Utilities Division abruptly became the focus of attention. Conflict was no longer primarily between individual segments of the petroleum industry but between the industry and its customers. For the first time, Commissioners found themselves the center of the state's attention, expected quite explicitly to act as defenders of the general public. Whether or not they succeeded, they had passed a turning point in the history of the Railroad Commission and in the politics of Texas.

BACKGROUND TO THE CRISIS

The 1950s and 1960s were decades of increasing prosperity for the natural gas industry. National consumption of gas, roughly six trillion cubic feet in 1950, topped twelve trillion in 1960 and totaled nearly twenty trillion in 1970.[27] During those years, the average price in Texas climbed from under five cents per thousand cubic feet to over fourteen cents.[28] In such an expanding market, aggressive business practices could yield large rewards. As new and established companies scrambled for customers, the industry acquired a reputation for wheeling and dealing that was frequently deserved.

There were three things a company had to attain in order to become successful in the natural gas industry. It needed a certificate of convenience from the Federal Power Commission, attesting to its viability as a supplier. It needed bank financing. It needed a buyer for its product. Getting all three of these was largely dependent on having adequate reserves of gas. The FPC required that companies show evidence of a twenty-year supply in order to be issued a certificate.[29] Banks demanded a similar stockpile before they would extend loans. And customers, particularly municipalities, insisted on assurances that their gas supply would remain stable for many years. The key to selling gas, therefore, lay in acquiring enough reserves to satisfy the FPC, banks, and potential customers.

Or, more accurately, the key to selling gas lay in *appearing* to acquire enough reserves. Because so much depended on the beliefs of others in one's gas supplies, there were great incentives for companies to overestimate their reserves. In fact, it was common knowledge in the gas industry in the 1960s that official company figures for reserves often bore little resemblance to anything those companies actually possessed underground.

That company estimates of their available gas were often inflated did not necessarily mean that they were lying. Estimating recoverable reserves of oil or gas is an extremely inexact science, and petroleum engineers with the purest motives can often arrive at quite different conclusions after studying the same evidence. When the inherent uncertainties of estimation were combined with the companies' desire for the largest possible reserves, it was only natural that the highest plausible estimate would often be accepted as authoritative.

This tendency to fictionalize reserve estimates caused little nervousness within the industry, because company executives were confident that they could acquire more gas in the future and so would never be caught short. In the vernacular of the industry, they "bet on the come"; that is, they gambled that there would always be enough gas available to be purchased so that their commitments would never outrun their supplies. Many gas companies therefore staked their business lives on a series of gambles. One company, however, took the tradition of betting on the future one step further. This was Coastal States Gas Corporation.

Coastal State's chief executive, Oscar Wyatt, was the most famous and flamboyant of the gas industry's gambling tycoons. In the mid 1950s, Wyatt had employed audacious, innovative, and (to his competitors) infuriating business techniques to secure a virtual monopoly over the transportation of gas in Central and South Texas.[30]

As the 1960s dawned, he turned his attention to establishing a monopoly over the distribution of gas in the lower third of the state.

Wyatt's method of acquiring this monopoly lay in his willingness to take an astonishing gamble. Coastal States and its various subsidiary companies offered to supply gas to Austin, San Antonio, Corpus Christi, and many smaller cities at what was even then a relatively low price—about twenty cents a thousand cubic feet. Furthermore, Coastal offered to sign contracts with the cities to continue to supply gas for twenty years at substantially the same price.

There was some sentiment on the Railroad Commission in 1962 to prevent the cities from signing these long-term, fixed-price contracts with Coastal. To the Commissioners, Oscar Wyatt's gamble did not look at all safe from the standpoint of his customers. But an order forbidding the contracts would have meant that the Commission was imposing higher prices on consumers. Even in 1962, that would have been politically risky. In addition, the Commissioners had their hands full that year with the slant-well scandal. So the Commission remained inert, and the cities were allowed, in effect, to take their own gamble on the reliability of Coastal States.

Later, much was made of the fact that, when these contracts were signed, Coastal did not in fact possess the gas reserves it claimed to have.[31] But this is not the essential point, for other gas companies frequently did not have sufficient reserves when they agreed to supply gas. Like Oscar Wyatt, they bet on the come. The difference between Coastal and other companies was that Wyatt gambled not only that he would find sufficient future gas but that he would find it at the same low price.

By the late 1960s, Coastal and its chief subsidiary, the Lo-Vaca Gathering Company, had become the only supplier for over four hundred customers in Central and South Texas, including the area's three large cities. It also supplied significant portions of the gas used in other sections of the state. Dallas and Fort Worth, for example, relied on Lo-Vaca for about 20 percent of their gas.[32] Besides its industrial customers, roughly four million people depended on the Lo-Vaca system for cooking and home heating. By the early 1970s, Coastal/Lo-Vaca was contractually obligated to deliver 1.9 billion cubic feet of gas a day, with estimated peak loads of 2.4 billion.[33]

This spectacular growth in its business made Coastal States the bluest of blue-chip stocks. In a very few years, the company had attained a market value of a billion dollars. It earned over 22 percent on stockholders' equity.[34] With a slogan of "every year a record year," its annual earnings were over 16 percent throughout the late 1960s.[35] Investors prized Coastal as a safe, dependable money-maker.

All this confidence would be justified as long as Coastal States could find enough cheap gas supplies to refurbish its inadequate reserves. And, all through the 1960s, Lo-Vaca's luck held good in this respect. But as the nation's gas supply dwindled, and as the price of the remaining discoveries began to climb, Oscar Wyatt's gamble failed.

As the ascending line of Lo-Vaca's contract commitments approached the descending curve of its deliverable reserves, in 1969 and 1970, the Coastal/Lo-Vaca system engaged in a series of business transactions that, not many years later, appeared breathtakingly irresponsible. First, Coastal/Lo-Vaca sold off some of its gas reserves, thus improving its short-run profit showing but, of course, further jeopardizing its long-run ability to meet its contracts. Second, Coastal banked gas from other suppliers by borrowing large quantities at the then prevailing price, under the condition that the supplier could withdraw an equivalent amount from the Lo-Vaca system at any time, with no upward price adjustment.[36] Like any other addicted gambler, Coastal/Lo-Vaca thus borrowed more and more heavily to finance ever greater wagers, hoping ever more desperately that the next spin of the wheel would yield the jackpot that would cover all past bets. And, like most compulsive gamblers, Lo-Vaca ran out of credit before its luck changed.

It was not that there was no more gas in Texas. But falling reserves had driven the price of gas from newly discovered fields beyond the point where Lo-Vaca could bid on it and hope for a profit. If Lo-Vaca bought gas at, say, twenty-five cents per thousand cubic feet and sold it at the contract price of twenty cents, not many years would elapse before it went bankrupt.

Trying to rescue themselves from this potentially fatal dilemma, the Lo-Vaca management in 1971 approached its customers and asked for a renegotiation of its contracts. Industrial customers were generally receptive to this request. But representatives of the major municipal customers, having to answer to their voting residents, demanded in return that they be given full information on Lo-Vaca's reserves and financial status. This Lo-Vaca was not prepared to grant, and nothing was done. So, in the winter of 1972–73, Lo-Vaca, and therefore four million Texans, began to run out of gas.

REGULATION AS BENIGN NEGLECT

While the gas industry in general and the Coastal States system in particular were thus constructing a political explosion, the Railroad Commission had been evolving from an agency mainly responsible for regulating the petroleum industry to one primarily devoted to representing that industry's views. As the oil surplus disappeared in the late 1960s and early 1970s, and as state allowables climbed toward 100 percent, the Commission's prorationing of Texas oil became of ever smaller consequence. When it no longer had its thumb on the national spigot, the Commission really was, as its press releases had always claimed, just a conservation agency, whose main official tasks were husbanding the remaining oil and protecting the environment. Unofficially, the agency was, as Commissioner Ben Ramsey stated in the mid 1960s, "industry's representative in state government."[37] The Commission's greatest political task was symbolic: it carried the industry view of potential policy to the capitol buildings in Austin and Washington.

The Commission that existed between the death of Ernest Thompson and the onset of the energy crisis bore many resemblances to the one that had been sitting just before the rise of the oil prorationing problem forty years earlier. Both agencies were rather sleepy vestiges of their once mighty selves—the 1930 Commission having declined from its railroad-fighting days, the 1972 Commission having fallen from its position of oil umpire. Both were staffed by men who had made their reputations in other institutions of Texas government and had never faced a real policy-making challenge on the Commission. Serving as Commissioners in 1972 were Ben Ramsey, a former lieutenant governor; Byron Tunnell, a former Speaker of the Texas House; and Jim Langdon, a former chief justice of the Court of Civil Appeals.[38] In both Commissions, the occupants of the office did not foresee the problem with which they were suddenly confronted, failed to deal with it competently, and made it worse through inaction before new Commission members finally resolved it.

Unlike the case in the 1930s, however, the later Commission had the regulatory authority to forestall the crisis but failed to utilize it. The Gas Utilities Division of the Railroad Commission was responsible for preventing precisely the kind of breakdown represented by the Lo-Vaca case. Commissioners, through this division, were supposed to regulate gas-gathering systems and their utility branches. The Commission was supposed to audit the gas companies, regulate their rates in the public interest, and see that they kept adequate reserves.

But the Commission's supervision of gas utilities was almost wholly theoretical. The statute under which the Railroad Commission had been granted the authority to fix gas utility rates mandated that companies be allowed to make a profit that was "a fair return upon the fair value of the property used."[39] In the controlling Texas Supreme Court decision (the Alvin case), the judges had been unable to devise a clear rule for discovering what a "fair value" rate was.[40] In the ambiguity left by the statute and the court, the natural sympathy of the Commissioners for the industry led them to grant rates that were, by national standards, almost confiscatory. Other regulatory agencies permitted rates that provided a 14 percent return to the stockholders. The Railroad Commission allowed rates that ensured an 8 percent return on the theoretical value of the utility's property. If the "fair value" rates were translated into "return on equity," this meant that the Commission permitted utilities to make a 33 percent return. In other words, the Railroad Commission allowed Texas utilities to charge their customers almost two and a half times what other state regulatory bodies permitted their utilities to charge.

But even this generous allowance is misleading, for in effect the Commission could not enforce its own rules. It did not have the staff. In 1972, the Gas Utilities Division employed eight people. These had to keep track of the 205 gas-gathering companies in Texas.[41] The systematic sleight of hand engaged in by gas companies in the 1960s could not have been detected by such a small staff, even if Commissioners had been interested in tightening the reins. As the Lo-Vaca gathering system drifted toward disaster, therefore, the regulatory agency given authority to protect the public from such exploitation lacked both the will and the organization to prevent, or even identify, the problem.

THERE AIN'T NO CURE FOR THE WINTERTIME BLUES

Late in 1972, Lo-Vaca became unable to supply the full demand for its gas and began curtailing deliveries. By January 1973, curtailments had reached the point where San Antonio, Austin, and the Lower Colorado River Authority had to resort to burning fuel oil, acquired at premium spot-market prices, to keep their electric generators turning, and the University of Texas at Austin was forced to close for a week to conserve fuel.[42] By the time spring warmth caused the demand for gas to drop in April, Lo-Vaca's customers had suffered thirteen curtailments totaling more than sixty-five days, and homes all across South and Central Texas had stayed warm only through frenetic scrambling on the part of municipal officials to find alternative sources of energy.[43]

As long as Lo-Vaca could not buy gas without losing money, the situation would only get worse. In March, therefore, the company requested the Railroad Commission to set aside its long-term contracts to supply cheap energy to Texans. The hearings began May 1. It was soon clear to everyone that the cities which depended on Lo-Vaca for fuel were in serious trouble. Lo-Vaca, it turned out, possessed only 54 percent of the gas reserves it was supposed to have, could supply only 1.5 billion cubic feet of the 1.9 billion it was contracted to deliver each day, and would be able to supply only 1.1 billion by November.[44]

Unable to imagine a way out of this mess, the Commissioners stalled, all through summer and into autumn. On September 15, Byron Tunnell resigned. Three days later, Governor Dolph Briscoe appointed one of his aides, Mack Wallace, a former East Texas district attorney, to replace Tunnell.

Meanwhile, in July, District Judge Charles Mathews had taken charge of Lo-Vaca, split the company legally from Coastal, and appointed a special supervisor-manager, Mills Cox, to direct its operations. A receiver was not appointed because, it was thought, that would signal bankruptcy to the financial world, creating even more supply problems.[45]

Working under the theory that the imperative problem was lack of gas, rather than prices, Cox immediately began to buy reserves for the Lo-Vaca system, essentially agreeing to pay any price that producers requested. He could not go very far with this, because, if the company were forced to buy gas at ever climbing prices and sell it at the 1962 contract prices, it would quickly go into bankruptcy. Commissioners Langdon and Ramsey did not have a solution to the Lo-Vaca problem, but they were not about to let the state's largest gas utility commit suicide this way. The best short-term expedient seemed to be, first, to allow Lo-Vaca to temporarily pass along its increased costs to its customers and, next, to urge the parties to negotiate new contracts. This would entail increased utility costs for consumers, but that seemed better than allowing them to freeze in the dark.

So, on September 27, 1973, the Railroad Commission issued an interim order suspending Lo-Vaca's contracts and permitting that utility to charge its customers its own weighted average costs for gas, plus five cents per thousand cubic feet.[46] Unfamiliar with the details of the situation and wary of its political implications, the new Commissioner, Mack Wallace, refused to sign this "100 percent flow-through" order.[47]

Under the interim order, Lo-Vaca continued to buy gas anywhere possible, at whatever price was asked. By the date of the order, Cox was buying gas at fifty-five cents per thousand cubic feet, more than twice the price Lo-Vaca could charge under its suspended contracts.[48] By the end of the year, the company had bid prices within Texas to ninety cents, and, by mid 1974, it was paying well over a dollar. Under the 100 percent flow-through provision, Lo-Vaca figured the price of this new gas into the costs it was already paying producers under its older contracts, added five cents per thousand cubic feet, and passed this weighted average cost along to its customers. And so, with the price of gas shooting ever upward, the citizens of Texas entered the winter of 1973–74.

When Lo-Vaca had interrupted gas deliveries the previous year, no actual harm was worked on residential consumers, for city officials managed to hustle enough gas to keep everyone warm. As this was a temporary situation, the spot purchases of gas did little to affect utility bills. But, under the September 27 order, as the price of gas to Lo-Vaca doubled and tripled, so did the price at the burner tip in homes and factories.

Arriving just before the Arab oil embargo, when citizens were apt to become angry at energy policy makers anyway, the Railroad Commission's 100 percent flow-through order created a surge of consumer resentment that gradually became a storm of protest. Employees of the Gas Utilities Division had seldom received a direct consumer complaint in the previous five years. Now they began taking dozens of phone calls and letters a day from irate citizens demanding that something be done about their utility bills. Texas Representative Ron Bird and U.S. Representative Henry Gonzalez of San Antonio launched a barrage of criticism at the Commission for alleged cruelty to their constituents.[49] There were tumultuous mass meetings and a protest march on the governor's mansion.[50]

While officials of the gas industry complained that the Railroad Commission was damaging business by favoring consumers, representatives of the AFL-CIO, Common Cause, and other consumer-oriented groups poured a steady stream of abuse on the Commission.[51] The San Antonio newspapers played up the conspiratorial overtones of the problem, informing their readers, with front-page headlines, of dark deeds being perpetrated in Austin.[52] Even more sober and industry-oriented publications like the *Houston Chronicle* editorialized that it was "plain that the Railroad Commission majority sees the Commission as an arm of the oil and gas industry. If the Railroad Commission does not take over protecting the public interest as a true utilities commission, its importance will wither."[53] For the first time, Railroad Commissioners were getting attention from the general public, and they plainly did not like it.

THE CRISIS IN SUSPENDED ANIMATION, 1973–1977

But there was little more that the Commissioners felt they could do. Lo-Vaca's customers unquestionably had valid contracts, but that was irrelevant. Without higher prices there would be no additional gas. If the Commission forced Lo-Vaca into bankruptcy, a period of turmoil would follow, in which the gas supply of South and Central Texas would be insecure. After a while, another company would take over Lo-Vaca's operations (although, at the time, no Texas gas utility expressed the slightest interest in assuming management of a bankrupt Lo-Vaca). A new company, however, would be forced to pay the prevailing rates for gas, so consumers would not be better off. Yet those same consumers were in a state of insurrection, and fatal consequences obviously awaited the politician who appeared to ignore their interests. It was a no-win situation.

The Commissioners, therefore, hoped that Lo-Vaca and its customers would get together and amicably renegotiate the contracts, thereby getting everyone off the hook. Jim Langdon, in particular, was tireless in privately urging this course on the principals.[54] In fact, talks continued all through this period. But by early 1974, in the midst of the first winter of the energy crisis, the atmosphere between Lo-Vaca and its customers was so poisoned that realistic negotiations were impossible. The Lo-Vaca management saw representatives of the customers as grandstanding demagogues, who were trying to make political capital out of the situation instead of solving it. The customers saw Lo-Vaca as a deceitful, irresponsible company, quite willing to trample on the public interests in its pursuit of profit. Instead of agreeing to levelheaded compromises, the parties sued each other. For four years, negotiations dragged on, lawsuits flew like confetti, the price of gas climbed to over two dollars per thousand cubic feet, and the Railroad Commission remained paralyzed. Commissioners held hearings, issued some ill-conceived or peripheral orders, and hoped for progress. The situation stagnated.

Meanwhile, Railroad Commissioners faced electoral campaigns. Mack Wallace, appointed in September 1973, had to survive a Democratic primary in May 1974, and Jim Langdon was up for renomination at the same time. Both won with surprising ease. Perhaps the place of the Commission in the energy crisis had not percolated into the general public awareness. Jim Langdon was relatively well known in Texas politics, having been elected twice before. The lesser-known Wallace had not voted for the 100 percent flow-through order. In addition, both had great financial advantages over their rivals. At any rate, the 1974 primary was not one in which the Lo-Vaca problem played a major role.

The 1976 election was a different story. Ben Ramsey chose not to seek reelection, and seven candidates entered the Democratic primary, hoping to replace him. All ran as more or less consumer-oriented candidates, and two, Terence O'Rourke and Lane Denton, based most of their campaign appeals upon attacks on the Commission's handling of the Lo-Vaca problem. As will be explained in chapter 8, Jon Newton won this election largely because he was able to convince both consumers and producers that he was friendly to their interests. The 1976 election marks the point in the history of the Railroad Commission at which Commissioners began to explicitly acknowledge their responsibilities to ordinary citizens. With the election of Newton, a new era began for the Commission.

THE POLITICS OF SUCCESS

When Ramsey and Langdon were a majority on the Commission, they represented the old political perspective. Mack Wallace, a younger man with a more flexible approach to public policy, was a minority. The result was inaction and drift. But, with the arrival of Newton in January 1977, the balance on the Commission shifted to one more appropriate to the changed political circumstances. In one of his first public statements after taking office, Newton made the heretofore unthinkable assertion that the Commission had been the captive of the oil and gas industry and would have to change.[55] Wallace and Newton, outvoting Langdon, began a series of orders to chip away at the revenues Lo-Vaca was making under the 100 percent flow-through arrangement.[56] The difference in tone and content of

Commission deliberations was felt immediately by Lo-Vaca's municipal customers, and they remarked upon it among themselves.

There were also organizational changes. In 1976, the state legislature had passed a law significantly modifying the "fair value" standard used by the Commission in setting utility rates and increasing the size of the division staff to over sixty people.[57] The Railroad Commission became a real utilities regulatory agency.

Still, the negotiations between Lo-Vaca and its customers crawled on. Through the summer and into the fall of 1977, the parties wrangled. There was a widespread perception at the time that both sides were stalling: Lo-Vaca because it was making satisfactory profits under the interim order, now four years old, the representatives of the municipalities because they dreaded presenting their citizens with contracts calling for unnervingly high gas rates. This perception was not necessarily accurate. Some observers close to the action maintain that by December of 1977 Lo-Vaca and its customers were very close to a compromise agreement and that, in a short period of time, they would have announced a plan to solve the problem of gas supply.

Whether or not this is true, Newton and Wallace had also run out of patience. On December 12, with Langdon once again in dissent, the Commission issued what has come to be known as the kamikaze order. Lo-Vaca was directed to honor all its old contracts and refund to its customers the $1.6 billion that it had collected under the interim order. There would be no further flow-throughs; the company had to buy gas at $2 per thousand cubic feet and sell it at a tenth that cost. The Commission, in other words, ordered Lo-Vaca to go bankrupt.[58]

The effect of this order on the negotiations is ambiguous. Those observers who believe that the parties were close to an agreement argue that the kamikaze order disrupted the negotiations and set the process of accommodation back by years. Those who believe that Lo-Vaca and its customers were stalling are more apt to argue that the bankruptcy order, by presenting both sides with the prospect of imminent disaster, finally made them talk seriously. At any rate, Lo-Vaca and most of its customers announced a preliminary agreement—really an agreement to agree at some future time—within three weeks after the kamikaze order, and Wallace and Newton promptly canceled it.[59]

However uncertain the effects of the December 12 order on the negotiations, there is no ambiguity about its effects on the relationship between the Railroad Commission and the public. The order was a political masterstroke. No one wanted bankruptcy, but bruised consumers needed a champion to make some aggressive move against an industry that seemed to be in league with the Arabs to pick their pockets. From all over Texas, editorials praised the agency as the defender of the common citizen.[60] City officials congratulated the Commissioners on a job well done.[61] An editorial cartoon in an Austin paper, referring to the previous day's first-ever victory for the Tampa Bay Buccaneers of the National Football League and to a recent motion-picture awards ceremony, exclaimed:

> The Day That Boggled The Mind!
> In a single 24-hour period in December, 1977,
> While Mercury was in conjunction with the Moon:
> The Tampa Bay Buccaneers won a football game;
> John Wayne presented an acting award to Jane Fonda;
> And (gasp) the Texas Railroad Commission made a decision in favor of the people!![62]

It was clear that the Railroad Commission had turned a corner in its relationship with its constituency.

THE PRICE OF REALITY

As with the rest of the Lo-Vaca problem, the tentative agreement was subject to delay, misunderstanding, and altercation, and it took almost two years before all the major parties accepted the settlement plan, as it was called. But on September 4, 1979, with natural gas selling at the wellhead for $2.29 per thousand cubic feet, the Railroad Commission issued an order approving this plan, thus ending the most important part of the most vexatious policy-making episode in the history of the Commission.[63]

Under the settlement plan, the Lo-Vaca Gathering Company ceased to exist, and a new organization, the Valero Energy Corporation, consisting of the old Lo-Vaca gas utility pipeline and extraction plant operations, plus a retail gas distribution system of Coastal States, was spun off from Coastal. Its customers own about 60 percent of the equity of the new company. Valero is permitted to charge 100 percent of the weighted average cost of its gas, plus ten cents per thousand cubic feet the first twelve months and fifteen cents thereafter. Part of the profit from Valero's utility sales, added to the profit it makes from transporting gas for other utilities, will be returned to its customers in the form of credits on their utility bills. Although it will no longer have any corporate connection to Valero, Coastal States will spend between $180 million and $230 million to search for gas fields for the new company.[64] Most of the customers dropped their lawsuits.

The consumers can therefore expect only a very modest easing of their gas and electric bills, but at least they will exercise some control over the utility that services them. Together with the new vigilance of the Gas Utilities Division of the Commission, this means that Texans will probably be spared a repetition of the shocks that were the end result of the structure of the gas industry of the 1950s and 1960s.

CONCLUSION: A NEW RAILROAD COMMISSION?

Prior to 1973, the Commission's relevant political constituency was composed almost entirely of members of the petroleum industry, especially its independent producing segment. Oil and gas were cheap and abundant, and Americans as a group were unconcerned with their regulation. The functions of the Commission were consequently obscure inside Texas and unknown outside the state, except to industry members and people whose profession brought the agency to their attention, such as economists. This did not mean that the political task of Commissioners was easy. As the previous chapters have demonstrated, the Commission has been at the center of political struggle ever since the East Texas strike of 1930. But these were largely battles fought among parts of the industry, and Commissioners were able to deal with them within relatively narrow boundaries.

The Lo-Vaca problem changed this. When natural gas became almost instantaneously expensive and subject to supply cutoffs, ordinary people, both within and without the state, became intensely interested in any governmental authority that had influence over energy. The Commission's role changed abruptly from industry representative and referee to watchdog of the public interests, and its political environment changed from one emphasizing personal contact, private negotiation, and substantive decisions to one encompassing mass protest, blinding publicity, and symbolic reassurances.

The politics of mass symbolism is not like the politics of industry adjustment that characterized Commission policy making before the Lo-Vaca watershed. Pre-1973 politics consisted largely of allocating access to markets among producers and pipelines. That is, the conflicts with which Commissioners had to deal were over *tangible* rewards. Policy making after Lo-Vaca, however, often consists of making the appropriate proclamations to reassure consumers that the Commission is a defender of their interests. That is, the conflicts with which Commissioners now have to deal tend to involve *intangible* beliefs and emotional states. Prior to 1976, campaigns for the Commission often involved charges by challengers that the incumbents had set production allowables so that some segment of the industry (usually the independents) was not receiving its fair share. Challengers attempted to mobilize a section of the industry away from a current Commissioner. Since 1976, campaigns for the Commission are more likely to involve an attempt to portray an incumbent as having an anticonsumer attitude. Pre-1976 campaigns were narrowly focused, but at least they dealt with definable policies that could be changed by making specific choices. The campaigns that began in 1976 dealt with much broader and more diffuse issues and involved an attempt by challengers not so much to mobilize interests as to galvanize mass emotions. Before Lo-Vaca, the politics of the Railroad Commission was a politics of distribution; after Lo-Vaca, it was a politics of image building.

The altered character of campaigns is not the only change wrought in Commission politics by the gas shortage of the 1970s. The agency's position in national politics changed quickly from one on the periphery to one much nearer the center. This was made clear by the debates over the Natural Gas Policy Act in Congress in 1977 and 1978, of which the gas situation in Texas was one of the principal issues.

This swift reorientation of Commission politics to the changed reality of the energy crisis was almost as painful for the men of the precrisis Commission as it was for the helpless consumers paying expanding chunks of their income for utility bills. The older members could not adjust, and the evolution of a new policy-making atmosphere on the Commission was accompanied by a rapid turnover among the Commissioners.

The lasting effects of this change are not yet clear. As chapters 7 and 8 will emphasize, the Commission is still very much a part of the petroleum industry, and no Commissioner of the late 1970s has failed to inform industry members of his loyalty to them. Those chapters will establish that there are good structural reasons for this continued association. But the Lo-Vaca problem, by making consumers directly and immediately part of a Commissioner's constituency, has created a far more complex political equation for the Railroad Commission to solve. The future of the Commission will undoubtedly be more difficult, more eventful, and less historically homogeneous than its past.

6. Policy and Its Consequences

You pays your money, and you takes your choice.—American folk saying

In the five decades since the Texas Railroad Commission became nationally important, sixteen men, representing a variety of personality types, have served as Commissioners. Some have been intelligent and some dim; some have been high-minded and some petty or venal; some have been dogmatic and some flexible; some have been abrasive and some diplomatic. Some Commissioners are remembered by those who knew them with respect, others with contempt.

Despite such differences, however, the Commission's policy making has remained relatively consistent. There are obvious threads of political concern that run through the public proclamations and retrievable private remarks of Commissioners since 1930. Sometimes the motives underlying Commission policy have been easily understandable, because Commissioners have stated their thinking for the record. On other subjects, Commissioners have been more reluctant to explain their purposes, but lines of consistent choices make inferences easy to draw. Since the onset of the energy crisis in 1973, new concerns have entered the calculations of Commissioners, without erasing their old purposes.

The seven policy-making episodes analyzed in the preceding chapters provide the basis for a summary set of generalizations about the chief historical policy concerns of the Commissioners. Their elaboration will be the subject of the first half of this chapter. But such a summary would be purposeless unless the general importance of the Railroad Commission can be established. In other words, it will be necessary to come to some tentative conclusions about the effects of Commission decisions on the price and availability of petroleum and about the desirability of those effects. The second half of this chapter will consequently be devoted to exploring the economic impact, on Texas and the nation, of the policies adopted by the Railroad Commission and to evaluating that impact.

PATTERNS OF POLICY MAKING

Both in their public and private statements and in the structure of their policy choices, the sixteen Commissioners have revealed consistent patterns of motivation and purpose. These patterns can be summarized with five generalizations.

First, for five decades, Commissioners have been concerned with the physical conservation of oil and gas and the protection of the environment. This interest has been observable at several levels. In terms of large-scale decisions, there is no doubt that the prorationing of crude oil served to prevent an enormous waste of this resource. Before prorationing, unmarketable oil ran down creeks, evaporated, accidentally burned, and, in general, was subject to many varieties of loss. There is no record of how much oil was frittered away in the unregulated industry, but it must have been substantial. Even more important, controlling production conserved reservoir energy, so that far higher proportions of the oil in place could be recovered. The historical record indisputably shows that prorationing saved most of the nation's oil reserves from despoliation.[1] Similarly, the campaign to eliminate gas flaring was a large-scale policy-making episode whose sole aim was physical conservation.

These are the more famous examples. But, in addition, for fifty years Commissioners have been making and enforcing regulations to protect the environment from individual carelessness and greed. These regulations, and the forms they require, now fill a good-sized book.[2] As the preceding chapters have stressed, particularly in regard to slant wells, the regulations have evolved in a direction making them ever more comprehensive and detailed. The substances involved in drilling, such as cement, must be of a certain type and applied a certain way. Storage facilities must be built and maintained to rigorous standards. Dry or depleted wells must be abandoned under specified conditions, taking detailed precautions to avoid hazards to humans or the environment. Pipeline companies must run their transportation networks so as to lose an ever diminishing proportion of their natural gas in transmission. And so on.

To enforce these regulations, the Commission maintains a staff of field inspectors who oversee the production, transportation, and storage of petroleum. These employees not infrequently force members of the industry to alter their operations. There are no favorites in this process; even famous members of the industry, who have contributed large amounts of money to Commissioners' campaigns, have been required to spend time and money to change their practices, because someone on the Commission staff thought that their mode of activity was a threat to the environment. The constant claim of Commission public relations releases that the agency is a vigorous conservation body is not a propaganda smoke screen—it is the plain truth.

This does not mean that the Commission has always been successful in defending the environment. As the slant-well scandal illustrated, good intentions are not enough when regulating a dispersed, resourceful industry that resists supervision. The Commission has occasionally been harshly criticized for serious lapses in its efforts to stop pollution.[3] There is undoubtedly room for improvement in its protection of Texas' land and water.

But this first aspect of policy making is the least interesting politically because it is the least controversial. Individuals who are ordered to clean up an outdoor storage facility or replug a well often protest, but there is no general, widespread opposition to physical conservation or antipollution measures from within the industry. The problem of protecting the environment is one of individual evasion, not collective defiance, and is therefore of secondary importance.

Second, and perhaps lying at the root of all their other concerns, it is clear that the Commissioners have in some sense seen themselves as managers of the Texas economy. As citizens of a southern state, Texans as a group share a history of poverty and exploitation. It has been oil and the by-products of oil, such as chemicals, that have pulled the state up into the industrial world until it is now an acknowledged economic power.[4] Commission policy must be understood as a series of measures designed to facilitate this process of economic development.

When Ernest Thompson ran for his first full term on the Commission in 1936, he gave speeches that aptly illustrate his position on the importance of the petroleum industry to the state economy and the duty of the Railroad Commission to shepherd that relationship:

> I think it is significant that more than a million persons in Texas directly derive the income upon which they subsist from the industries preserved and carefully controlled by the Railroad Commission of Texas. . . . During the past four years Texas has drilled and brought into production 43,000 new oil wells. These wells would never have been within the realm of possibility had it not been for the stability of the oil industry

> made possible by . . . the Railroad Commission. Out of this drilling activity, 15,000 man-days of work have been created and more than $75,000,000 have been made available to labor. . . . The price of crude oil early in my administration was restored to a dollar a barrel and has existed steadily at that point for three years. Oil producers and refiners are prosperous and contented. Into the treasury of the state flows a constant cash revenue—a revenue that wipes out deficits, that balances budgets, that builds highways, and that affords the funds necessary for public education.[5]

In 1967, when Jim Langdon sought to explain the alliance between the industry and the Commission, he demonstrated that his views were identical to Thompson's:

> The role of the Commission as a state agency charged with the administration of a public function for the benefit of the state as a whole, certainly includes that of participating as actively as possible in the encouragement and development of a vigorous, healthy, prosperous oil and gas industry. Moreover, since the petroleum industry is not only the most valuable source of state revenue but is also one of the state's best and greatest employers, it behooves us to keep it prosperous and healthy.[6]

Some Commission policies that otherwise appear rather perverse make simple sense if they are perceived as stemming from a conscious desire on the part of Commissioners to manipulate the petroleum industry to bring general prosperity to Texas. The best example is the series of decisions on well spacing and allocation described in chapters 3 and 4. Commissioners were for decades adamant in giving every landowner, no matter how insignificant, the right to drill a well and an allowable large enough to make it profitable. The result was hundreds of thousands of economically unnecessary wells, a bias toward inefficient producers, and the nurturing of a state infrastructure of drillers, work-over services, pipelines, and all the other industrial paraphernalia of the oil fields. Another result was a dispersal of drilling, so that, even in the late 1970s, approximately 650,000 Texans received monthly royalty checks.[7]

Economists have expended much ink demonstrating that this policy was irrational. One called it "a featherbedding scheme."[8] And so it was, looked at from the viewpoint of economists: rationality results from getting the most oil for the smallest expenditure of social resources.

But that is not the perspective adopted by Commissioners, and this brings us to their third general policy-making motivation. To understand the concerns of Commissioners, we must imagine a Commission run by economists. What would be the result of a spacing and allocation policy that took only "optimal return on investment" as its guiding beacon? There would be far fewer wells, and those that did exist would be the most efficient producers (for example, those capable of producing the most oil fastest, because they were in the richest part of the richest fields). And who owns these wells? The major integrated companies, because they, being wealthier and better able to support a large research staff, have appropriated the best land. And who owns the major oil companies? Northern capital. So who benefits from an economically "rational" spacing and allocation policy? Yankees. And who benefits from an "irrational" policy? Texans.

In order to help the Texas economy, therefore, Commissioners adopted a policy of favoring independent producers and small landowners at the expense of the major companies. The smaller the producers and landowners, the more they benefited from Commission policies. If we accept that it is politically rational for Texas politicians to favor Texans, then it is clear that political rationality mandated economic irrationality.

There was nothing indirect or embarrassed about this policy. In public statements over many years, Commissioners have pointed out the colonial relationship that has existed historically between Texas and the Northeast and suggested that it was their task to alter it.[9] By favoring independents and the smaller landowners, the Commission in effect taxed the major companies and, through them, northern consumers, for the benefit of Texans.

Other Commission policies besides those pertaining to spacing and allocation had the effect of protecting Texas independents from market forces that would otherwise have worked to the advantage of the major companies. The protection of correlative rights, a complex of policies addressed to the problem of ensuring equitable access to markets, basically involved measures to shield independents from the economies of scale, ownership of pipelines, and, in general, the superior resources of the majors. Across-the-board statewide prorationing forced the majors to connect to independent-dominated fields, instead of merely taking more from their own. Vigorous enforcement of common purchaser and common carrier statutes ensured every producer a market, thus further protecting independents. Exemption of marginal wells from prorationing underwrote their survival in an economic situation that otherwise would have resulted in their premature abandonment.

It is possible to test the interpretation that an important function of the Railroad Commission has been to protect independents from market forces. If the present argument is correct, then without the Railroad Commission independents would have prospered in economic good times for the domestic industry and failed in economic hard times. That is, in the late 1940s and early 1950s, independents would have been a significant part of the state industry, but in the late 1950s and the 1960s they would have been squeezed out at a much more rapid rate than actually was the case. Then, in the middle and late 1970s, they should have prospered again.

A pure test of this hypothesis is impossible, for of course there are not two historical Texases, one with and one without a Railroad Commission. But it is plausible to employ California as a surrogate. Although it has a private voluntary-proration committee and has a state agency charged with enforcing conservation, California has never had an equivalent of the Railroad Commission, with governmental authority to prorate to market demand, enforce the protection of correlative rights, and so on. In California, independents are on their own. Yet California has been a major producing state for almost as long as Texas. There are differences of detail, of course: California tends to have smaller reservoirs in more complex geological formations, filled with heavier-grade oil. But these are differences of degree, not of kind. California will be used here as the best example of an important producing state in which the independents are not protected.

If the present interpretation of the Railroad Commission is sound, independents should thrive in California until about 1958, after which they should steadily lose ground to the majors. They should revive slightly in 1967, when an Arab-Israeli war temporarily halted oil production in the Middle East, thereby greatly improving, for a short period, the market for American oil. After 1967, however, they should quickly decline. Then, beginning in 1973, when any drop of oil became instantly valuable, they should again become prosperous. During these changes, however, Texas independents should retain a relatively strong position in that state's industry.

The best way to test this possibility would be to chart the independents' percentage of California's production against the independents' share of Texas' production over time. Unfortunately, this is impossible. Production data by company are available for California back to 1940 but for Texas only back to 1961.[10] Furthermore, the Texas data are based on a one month per year's compilation, the California data on an entire year's production. For these reasons, no direct test of the hypothesis can be attempted.

An indirect test is possible, however. The overwhelming majority of marginal wells are owned by independents. Major companies usually do not wish to be bothered with such low-yield investments. The best industry estimate available indicates that 80 percent of stripper wells are the property of independent producers.[11] The Interstate Oil Compact Commission keeps records of stripper wells by state. By using this data, we should be able to compare the fates of marginal wells in California and Texas over the last three decades and thereby get some notion of the comparative position of independents in the two states.

Strippers are exempt from prorationing by Texas state law, so any effect on their economic position cannot be completely ascribed to the actions of the Railroad Commission. But the Commission enforces the legislative mandate; moreover, pipeline connection orders, ratable-take directives, and many other protective actions are the result of Commission decisions. Whatever effect there is to be discovered from public policy toward marginal wells, therefore, is to be credited mostly to the Commission.

Figure 1 shows the proportion of total wells in California and Texas that were marginal from 1950 to 1976. The figure makes it immediately clear that strippers in California have formed a markedly varying proportion of the total number of wells in that state. The proportion of strippers expands when times are relatively good and contracts when times are bad. In Texas, however, strippers constitute a much more stable proportion of the number of wells.

Because it is indirect, this is an inconclusive test, but it does vigorously support the hypothesis: the Railroad Commission has successfully pursued a policy of buffering independents from market forces within the state.

The fourth pattern of policy making has been more or less a byproduct of the two previous patterns. Because they have focused on improving the position of the Texas industry, Commissioners have inevitably favored the interests of producers, as a group, over the interests of consumers, as a group. Prorationing, perhaps, and spacing and allocation, certainly, have caused Texas oil to be more expensive than it would have been in the absence of these policies.

Figure 1. Marginal Wells as a Percentage of Producing Wells, California and Texas, 1950–1976

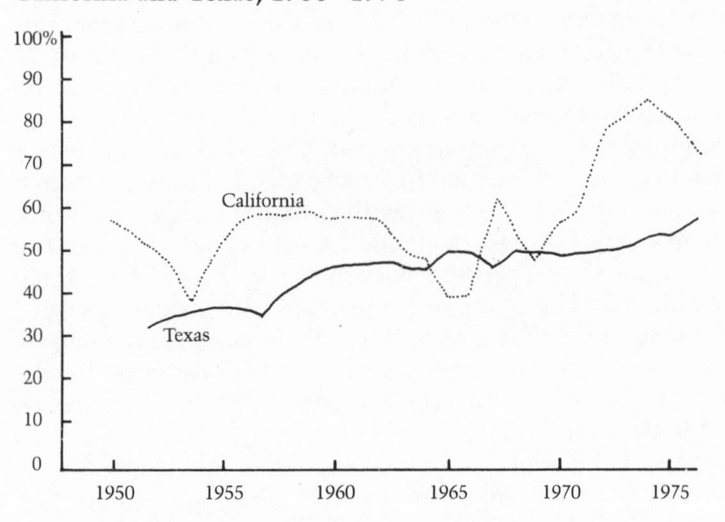

SOURCES: National Stripper Well Survey for relevant years, supplied by the Interstate Oil Compact Commission; *Texas Energy: A Twenty-Five Year History* (Austin: Governor's Energy Advisory Council, 1977), p. 40; Conservation Committee of California Oil Producers.

Fifth and last, however, the onset of the energy crisis has somewhat altered these traditional patterns of policy making. The Lo-Vaca problem focused attention on the gas utilities responsibility of the Commission, and the subsidence of that particular problem has not caused this attention to diminish. Now, whenever the Commission makes a significant decision in the gas utilities field, it is news.[12] Moreover, recent elections have witnessed a series of self-proclaimed consumer candidates challenging incumbent Commissioners. As politicians, the Commissioners are well aware that acquiring the label of anticonsumer could mean their political death. In response to these changed circumstances, they have altered the tone and content of their public statements and substantive decisions to become more consumer-oriented in an attempt to prevent a successful challenge to their seats.

The energy crisis has thus expanded the Commission's constituency and caused a shift in some Commission policies. The change has so far not been overlarge, not nearly enough to justify the expectation of a new direction for the agency. The times are unpredictable, however, and future elections could conceivably bring in new Commissioners who defined themselves in opposition to the interests that have traditionally dominated the agency. If that happens, patterns of choice among Commissioners could change dramatically.

CONSEQUENCES

The main outlines of policy making on the Commission are now relatively clear. The consequences of that pattern within Texas are also fairly clear: prosperity has been shared more equally. But a more important issue deals with the impact of Commission policies beyond the state's boundaries. What was the effect of the Commissioners' political choices on the availability and price of petroleum in the consuming states? What would the country have been like without the Texas Railroad Commission? It is not possible to answer such questions with precision. The data are too fragmentary. Some general conclusions can be attempted, but no one should think that they are more than rough approximations of reality.

In order to discuss the national costs of Commission policies, it will be necessary to rely heavily on the published work of economists. They would freely admit that their conclusions are tentative, so ours must be doubly so. What follows is to be taken as suggestions, not as definitive answers.

To begin with, there must be some accounting of the price of market-demand prorationing. If the Railroad Commission had contented itself with relegating each oil field's production to its maximum efficient rate (MER), there would have been no controversy over prorationing. That policy would have been no more than a conservation measure. For the convenience of the petroleum industry, however, the Commission, until 1972, suppressed each field's production so that the total from all fields just equaled the current demand for the state's crude at the current price. For forty years, the total dictated by the market demand was well below the amount that would have been produced if all fields had been permitted to flow at 100 percent of MER. Actually, as explained in chapter 3, MER was not even used for most fields. Instead, a yardstick based on the depth and acreage of a field was employed to set allowables. But, with permissible production held so low so much of the time, whether MER or yardstick set the theoretical maximum was largely irrelevant.

When the production totals of the market-demand proration

states were coordinated, the effect was to place a floor under the price of crude oil. Production up to MER would have flooded the market, forcing prices down in the short run. Although Commissioners have, throughout the years, issued many disingenuous statements denying that they ever intended to fix prices, they were far too intelligent not to understand that the by-product of market-demand prorationing must be price support. The real question is not whether their policies affected prices but whether the short-run price-supporting consequences of prorationing were counterbalanced by other, long-run factors.

Economist Erich Zimmermann explored this question in the 1950s. Examining historical price and supply data, he demonstrated that, in the days before prorationing, both displayed something of a zigzag pattern. A period after the discovery of a major field would be followed by a drop in exploration and a decline in price as the new crude supply inundated the market. After an interval of time, as the new field was depleted and there were no further discoveries, supplies began to tighten and the price shot up. The higher price led to renewed exploration, which led to more discoveries, which again lowered the price, and so on.

Zimmermann argued from this pattern that prorationing served to lower prices in the long run. By guaranteeing a market for new crude, it both encouraged oil companies to explore for more reservoirs and maintained a buffer of excess productive capacity. If there had been no prorationing, exploration would have again occurred only during times of shortage, and the consumer would have again been the victim of oscillating prices. By stabilizing the supply, prorationing made for lower prices in the long run.[13]

This is a powerful argument and, in the absence of complicating historical factors, it might be persuasive. But the prorationing system in Texas was intertwined with a plethora of other measures that were connected to it not logically but politically. Protection of marginal wells, narrow well spacing, Commission pressures on pipelines to connect to geographically remote fields, and all the other trappings of the system were not strictly necessary to market-demand proration. Each was designed to protect small producers and landowners from the debilitating effects of the free market. Each added economic irrationality to oil production, thus raising costs, thus requiring higher prices to sustain profits. The effects of prorationing as a pure process cannot be separated from its accompanying baggage, at least in Texas. Zimmermann's argument about the long-run beneficial effects of prorationing is therefore largely irrelevant. The consumer was paying not for prorationing only but for a system of production control. Except for Zimmermann, economists are united in arguing that this system raised both the short-run and the long-run price of oil.[14]

The cost of prorationing to the consumer is consequently not calculable. But the cost of the system of production control is. In 1965 another economist, Morris Adelman, published some estimates of these costs based on the best available industry data. Using a profusion of statistics, he concluded that the national price of supporting an inefficient production apparatus was roughly four billion dollars per year.[15] Adelman emphasized that the data were shaky, but no subsequent efforts have been made with more reliable information, and economists have frequently accepted his conclusions as authoritative.[16] His figures will be used here as the basis of some rough estimations about the cost of the Commission's system of production control to the ordinary consumer.

Assuming for the sake of simplicity that Texas produced half of the nation's prorated oil in the 1950s and 1960s, and further assuming that Texas' contribution to the national cost overrun was equivalent to its share of production, then the national price tag of the system of production control in the Lone Star State was about two billion dollars a year. We also assume that this extra cost was passed on to the consumer through higher retail prices. The problem becomes simply one of calculating the effect of the extra two billion dollars per unit of petroleum product consumed.

In the early 1960s, about 46 percent of the nation's crude oil was refined into gasoline. Assuming that the cost of production control in Texas was distributed evenly among products at the retail level, then roughly 920 million dollars was thereby passed on to consumers at the gas pump. The United States used approximately sixty-four billion gallons of gasoline a year in that era, at a retail cost of about 31 cents a gallon.[17] Nine hundred twenty million dollars, distributed among sixty-four billion gallons, comes to about 1.4 cents a gallon.

As an imprecise estimate, then, the Railroad Commission's production control system cost American consumers about a cent and a half a gallon every time they bought gasoline in the early 1960s. As a comparison, state gasoline taxes averaged six cents a gallon in that period, and federal taxes averaged four cents a gallon.[18]

But slightly higher consumer prices are not the only potential cost of inefficient regulations. It has already been explained that crude-oil imports were such a threat to domestic producers because they were so radically inexpensive. The crucial questions are these: did the system of production control, of which the Railroad Commission was the principal component, *cause* the import problem by making American oil so costly? Or would foreign oil have been cheaper and hence more desirable than American crude, regardless of any possible movement toward more economical production in the United States?

The figures on this point are again more ambiguous than we would like, but they are sufficiently clear to enable us to reject the notion that the Commission and its related agencies somehow caused the import problem by raising the cost of producing American oil to uncompetitive heights. Foreign oil was extremely inexpensive primarily because it was frequently found in huge reservoirs and, therefore, was amazingly cheap to produce. As explained in chapter 2, when the East Texas field was discovered in 1930, its capacity of about 5.5 billion barrels was thought to be immense. Fifty years later, it is still the largest known field on the continent, outside of Alaska. By 1969, however, East Texas was only the twenty-fourth-largest known field on the planet. The Burgan field in Kuwait contained 62 billion barrels, the Ghawar field in Saudi Arabia 45 billion, and the Bolivar Coastal field in Venezuela 30 billion.[19]

With these and other supergiant fields at their disposal, the major oil companies could produce and transport crude oil to their refineries in the United States so inexpensively that most American oil could not have competed, no matter what changes were made in the American structure of production. Adelman, a vociferous critic of the entire domestic control system, estimated that by eliminating all the irrationalities in that system U.S. oil production costs could be made to fall by about $.85 a barrel. He also calculated, however, that foreign crude arrived in American refineries costing $1.25 a barrel less than domestic crude.[20] If these figures are roughly accurate, then, even if the Railroad Commission and the other state agencies had uprooted the domestic system completely and replaced it with one based exclusively on efficient production practices (which would have nearly wiped out the domestic independents), U.S. oil would still have been about $.40 a barrel more expensive than foreign oil. The Railroad Commission can therefore not be blamed for the national import problem.

EVALUATION

For fifty years, certain economists, political scientists, and journalists have attacked the Texas Railroad Commission as an enemy of the consumer and a threat to national security. For the same length of time, different economists, political scientists, journalists, members of the industry, and the Commissioners themselves have defended the agency as a guardian of the public interests. As usual, the truth seems to lie between the two extremes. But, from the perspective of the 1980s, the Commission's faults seem less grave and its virtues seem more substantial than we might expect.

There seems to be little doubt that the Commission has successfully enforced practices that have encouraged the conservation of petroleum. Before prorationing, producers were lucky to get out 10 percent of the oil originally in place in a reservoir. Today, industry experts anticipate that 83 percent of the original oil in the East Texas field, for one, will eventually be recovered. Most of this oil, of course, was not produced and will not be produced at the optimal level of economic efficiency. But given the technological ignorance with which it had to work for many years, and given the intense tugging of interests that surrounded it, the fact that the Commission was able to impose such responsible practices on the industry is a genuine, praiseworthy accomplishment.

As already explained, by favoring small independents and landowners in its regulations, the Commission raised the price of domestic oil and, hence, of gasoline, heating oil, and so on. But, for a variety of reasons, it is hard to condemn this practice as an assault on the ordinary American citizen. In the first place, the small "tax" put on retail products by the Commission hardly seems consequential when measured against the escalating extortions of OPEC in the 1970s.

In the second place, there is some doubt whether the price of petroleum was ever too high. All during the 1950s and 1960s, the retail cost of petroleum products rose at a lower rate than the consumer price index.[21] In other words, the real cost of petroleum, after taking account of inflation, fell steadily. Oil may have been more expensive than it would have been in the free market, but it was still a relative bargain.

Additionally, some economists now argue that the United States should never have allowed its petroleum prices to be so low. Cheap gasoline, they contend, led to urban sprawl, because citizens, anticipating eternally inexpensive transportation, could live many miles away from their workplace. For the same reason, rapid transit systems died stillborn for lack of funding. In contrast, Europeans taxed gasoline so heavily that their cities remained compact. Thus, when the 1973 oil boycott hit, European commuters were better able to cope. The argument is that the United States should all along have been taxing gasoline so that its greater expense would make Americans drive less. In other words, the historical price of American oil has been foolishly low.[22] If this is true, it makes no sense to attack the Railroad Commission for keeping the cost of petroleum too high.

In the third place, no critic of the Commission has explained why the resources of Texas should be exploited for the convenience of out-of-state consumers. Texans perceived the petroleum "tax" as merely returning to the state some of the wealth that Yankee railroads, banks, and so on had confiscated over the years. Given the post–Civil War history of the United States, it is difficult to fault this attitude.

Moreover, much of the disapproval of the Commission's method of taxing out-of-state consumers seems to stem from the fact that it was indirect and deceptively packaged. If the state legislature had imposed an even much higher tax on petroleum, however, it would probably have been perceived as a legitimate state power. It is true that the Commission was not honest in its insistence that the production system was entirely a conservation measure, but this does not mean that its policies were predatory. The Commission was somewhat more roundabout in its taxing than the legislature would have been, but the results were possibly more beneficial to the state, because they sustained an industry. Through its regulations, the Commission taxed consumers in a way that redistributed wealth from non-Texans to Texans in a more efficient manner than a government program could be expected to accomplish. Given the relative mildness of the "tax," it hardly seems to be the expropriation that it has often been portrayed to be.

An additional consideration of the Railroad Commission as a policy-making body concerns the value to the country of supporting an immense number of relatively inefficient independent oil companies with favorable regulations. This topic will be addressed in chapter 9.

Finally, there is the problem of national security. From 1941 on, every defense of the prorationing system in general, and of Commission policies in particular, contained a variation on the following argument: during times of crisis, the nation needs a large reserve productive capacity. The Allies won World War Two because they were able to expand oil production to almost any desired level; the Axis failed because it ran out of petroleum. During the Middle East wars of 1956 and 1967, the reserve capacity of the United States prevented Arabs from using the threat of an oil embargo to apply political pressure on Western Europe. Without such a reserve, not only Europe but the United States itself will forfeit their foreign policy–making independence to whatever governments control their oil supply. By guaranteeing a share of the market to every well in every field, and by supporting the price, the prorationing system encourages exploration for reserves far in excess of what is needed immediately, thereby maintaining that extra capacity. Therefore, prorationing is vital to the security of the United States.

In light of history since 1973, this argument is compelling. But it has been rejected by some economists, and their views must be considered. Pointing to the fact that the excess capacity disappeared by 1972 despite prorationing, they draw the conclusion that the system did not supply in practice the incentives to explore that it promised in theory. In fact, they contend that the production control system actually discouraged exploration. This argument, as stated by Adelman, is worth quoting:

> Why risk money . . . to find new oil which, after a brief grace period, must be produced at a fraction of well capacity? The solicitude for strippers and other high-cost sources is killing the domestic oil finding industry. . . . If other things were equal, a higher price would mean greater inputs into exploration and development, but in the United States the higher price is secured only by greater restriction of output, which is a disincentive to higher inputs.[23]

Partisans of the control system would counter this argument by maintaining that the system did, in fact, contain incentives for exploration but that they were gradually overcome by the competition of imports. With domestic producers unable to sell much of their oil because of the availability of cheap foreign crude, they naturally would not drill for more, proration or no proration. The national reserve capacity declined despite the control apparatus, not because of it. Without the domestic system, reserves would have been exhausted many years earlier.[24]

This disagreement rests on an empirical question. As such, it is testable. One study has already addressed an important facet of the question. In 1964, Franklin Fisher published an econometric investigation of the relationship between the price of crude oil and domestic exploration. Fisher studied wildcatting in the years 1947 to 1955. His model suggested that, in that period, a 10 percent increase in the price of crude would induce a 28.5 percent increase in exploratory drilling and a 3 percent increase in the discovery of new reserves.[25] In other words, the higher the price, the more oil would be discovered. As an important function of the domestic control system was to support prices, this bolsters the notion that the Commission protected national security by helping create a petroleum reserve.

But the full story is not that simple. Fisher could not measure the effect of imports, since they did not begin to seriously undermine the domestic industry until after his investigation. His data suggest, however, that, in a context of increased production controls due to an import-depressed market, price supports would not encourage exploration. There seemed to be a threshold effect: slight suppression of production would encourage exploration, but heavy suppression would discourage it. If wildcatters know that they will be guaranteed a market, at the price of a small restraint on production, they will explore for more fields. If, however, they know that the price will be a heavy suppression of production, they will not explore. In fact, in a depressed market, a major consequence of the system fashioned by the Railroad Commission would be to shift exploratory drilling outside the state.[26] If this is so, it means that, after imports became heavy, Commission regulations actually damaged national security by deterring the search for new reserves within Texas.

There is ammunition for both critics and defenders of the Commission in these findings. If Congress had been persuaded to limit imports more significantly, as Commissioners constantly urged, then the production control system would have acted to increase the domestic reserves by motivating exploration. Given the actual reluctance of Congress to take such a measure, however, it seems that the control system was not in the best interests of national security.

An evaluation of the Railroad Commission as defender of the country's energy independence thus becomes impossible. With few imports, the Commission is defensible; with many imports, it is not. Critics of the Commission can plausibly argue that the nation would have been better off with high levels of imports and fewer domestic restrictions. Its defenders can just as plausibly argue that the country would have been better off with no imports and a retention of restrictions. The national security argument must be counted a draw.

NONPOLICY

This chapter so far has discussed Railroad Commission policy, the consistent choices of Commissioners over the decades. But it would not be complete without reference to nonpolicy, to the Commissioners' choice not to choose. Political scientists have long recognized that nondecisions are often as important as decisions in making public policy. The refusal of a public authority to consider that certain subjects are within its jurisdiction has as much effect on policy as do actual choices among alternatives on the agenda.[27] The Commission's nondecisions have not so far been explored in this book, because it is very difficult to study a void. But an account of the Commission would be distorted without at least a passing reference to its nonpolicies.

The most important nondecision the Commission has made over five decades has been its refusal to consider the problem of compulsory unitization. As explained in chapter 3, the only rational way to develop a petroleum field is by viewing it as an entity and making decisions about its exploitation on the basis of the most efficient application of technology to its entire structure. The pattern of multiple private ownership that is normal to the United States, however, leads individuals to ignore the most efficient plan of production while they concentrate on increasing their private profit. Over the years, billions of dollars have been spent needlessly, and millions of barrels of oil and cubic feet of gas have been lost, because individual profit rather than efficient overall recovery was the guiding principle of the Texas industry.[28] As individual interests cannot be expected to prevent this waste by endorsing a comprehensive view of a field, it must be imposed on them by some outside force. The most logical agency for this imposition is a government commission.

Since the 1940s, observers of the petroleum industry who desired its long-run best interests have been advocating that some authority in Texas compel the unitization of its fields.[29] The implementation of this policy would be fairly simple: the Commission would manipulate well spacing and allowable rules so that it was to the advantage of operators to unitize.

A policy of mandating unitization would of course have involved a modification of Commission rules favoring marginal wells, small landowners, and so on. But protective clauses could have been written into the rules to ensure that the industry's little people would not have been victimized by such a change in production practices. Additionally, compulsory unitization would not have altered the Commission's policy of favoring the smaller, independent-dominated fields in its statewide allocation formulas.

But the mere whisper of the phrase "compulsory unitization" around many independent producers or Texas landowners is sufficient to provoke opposition of a most violent and vocal sort. Compulsory unitization by definition involves overriding property rights and thus offends both the narrow self-interest of individual operators and the ideology of Texans in general.[30] Although there has been some sentiment on the Commission in favor of imposing unitization, the opposition has been sufficiently intense to prevent Commissioners from even considering it as a potential course of action. They have instead, as a group, denied that they have the authority to deal with the question and referred it to the state legislature. But the legislature has on numerous occasions demonstrated that it is no more capable than the Commission of braving the political opposition created by the possibility of compulsory unitization. As a result, Texas is the only major producing state without a compulsory unitization law.

The closest that the legislature and the Commission have come to endorsing the concept is by passing and encouraging adherence to a voluntary-unitization law. This statute has had some success: in 1976, Yates, the second-largest field in the state, was finally unitized, half a century after its discovery.[31] But most of the Texas fields are not unitized, and the waste continues.

Against the billions of barrels of oil that the Commission has preserved with conservation rules and prorationing must be set the millions of barrels that have been lost to future generations because of nonunitized production. Although most of the criticisms of the Commission's policies seem unfounded or overharsh, therefore, there is legitimate reason to condemn its most important nonpolicy.

THE FUTURE

These are the major trends and consequences of policy making on the Railroad Commission. But what of the future? Most of the importance of the Commissioners' choices about oil production lies in the past; the future promises policy battles over natural gas and, perhaps, over such infant state energy industries as lignite coal mining and the extraction of geothermal power. It would not be a surprise to discover that the coming decisions of Commissioners about these new fuels fall clearly into patterns established decades ago.

Yet the historical context of the Commission is changing, and it would be unwise to anticipate an uneventful policy future. All prospective Commissioners can be expected to cast themselves, like their predecessors, as managers of the Texas economy. All future Commissioners must likewise be expected to be supportive of physical conservation. But it is not impossible to imagine consumer protection candidates winning seats on the Commission in the near future. Such an eventuality would produce a new relationship between the Commission and the petroleum industry. The specific policy changes that would come from this altered relationship are impossible to predict, but they might be substantial.

When patterns of policy making have been examined, a question about what sort of political environment could have produced such consistent choices naturally poses itself. The next two chapters will attempt to answer that question by exploring the political influences on the Commission.

Part 3

Politics

7. Influence

They that govern the most make the least noise.—John Selden, *Table-Talk*

Considering the many conflicts that have centered around the Railroad Commission, the agency is held in remarkably high esteem by members of the petroleum industry. As the previous chapters have demonstrated, Commissioners have been responsible for allocating billions of dollars away from some segments of the industry and toward others. It has been deeply involved in some of the most divisive arguments in the industry's history. Given this tradition of partisanship, it is curious that virtually no members of the industry are willing to criticize the Commission on other than light and transient grounds. Individuals occasionally claim that their interests have been slighted by a particular decision, and there is some complaint that certain Commissioners are obtuse, inadequately trained, or the like. But there is no general opposition to the historical role of the Commission, and no general resentment has been aimed at any Commissioner who has served since the early 1940s.

This remarkable comity between the Commission and the industry will be explored in the present chapter. The following account of the relationship between the Commissioners and people in the oil and gas business is based on several dozen unstructured interviews. No conclusive findings are reported because the interpretations cannot be tested systematically. In the next chapter, however, quantitative election data will be introduced that lend support to the present argument.

Railroad Commissioners and members of the industry engage in a series of mutual applications of influence that mute conflict and preserve harmony. Furthermore, members of the industry interact with one another in a manner that suppresses hostility and creates consensus. The Railroad Commission has to be thought of not as a discrete governmental body outside the industry but as an integral part of that industry. This means not that the Commission is a "captive agency"[1] but that the Commissioners and the members of the industry have a reciprocally dependent relationship.

The application of influence can be studied at two levels, the personal and the organizational. Both will be described here, followed by a specific example of their relevance: gubernatorial appointments.

PERSONAL INFLUENCE

Much of the political science literature about the relationship between private power and public authority has addressed the subject of access. Large campaign contributions, possession of useful information, organizational strength, and similar resources are said to give some interests an in with policy makers. Personal contact is highly persuasive, and people with access habitually wield more influence than people without it. Consequently, much of the effort of interest groups and private individuals in the policy process consists of attempts to maximize their access to decision makers.[2]

This proposition is apparently a good description of politics in Washington, where the competition for influence is intense. It was until recently largely irrelevant to a study of the Railroad Commission, however. There was not a problem of access to the Commission until the mid 1970s. Anyone could see a Commissioner at almost any time. In the days prior to the energy crisis, when the life of energy regulators was slower, people could simply walk into their offices in Austin and probably be received. No one had to try to maximize channels of access to the Commission; they were open to everyone.

This accessibility of the Commissioners to personal visits was of great psychic importance to the smaller independent operators. In general, the life of an independent oilman is a relatively anxious one. Most independents run small concerns, taking considerable risks and never getting rich.[3] In virtually every resource—equipment, information, scientific expertise, ability to raise investment capital—the typical independent is far inferior to the major integrated companies.

Much of the contact that members of the industry have with the Commission consists of participation in hearings concerning the production rules for fields. In these hearings, company geologists, engineers, and lawyers, or the operators themselves, make presentations to Commission examiners and sometimes to the Commissioners about the physical properties of the field. As the production rules depend upon the Commissioners' findings of facts about the circumstances of the field, the ability to make a persuasive presentation can significantly affect the final rules and hence the allocation of production within the field. Because petroleum engineering is an inexact science, there is a fair margin of error in Commission determinations. The universal nightmare of small independent producers is that the majors, because their greater technical staff enables them to put on a better show at Commission hearings, will benefit more from Commission rules.

By leaving their doors open to personal visits from people in the industry, the Commissioners were able to allay these fears. Independents generally believed that, as long as they could present their side to the Commissioners, the organizational advantages of the majors could be neutralized. This was not a question of the independents' feeling that they must apply pressure to the agency to ensure a fair hearing; the probity of the Commissioners was unquestioned by both independents and majors. Instead, it involved the independents' perception that a fair hearing for them consisted not so much in a technical demonstration as in a chance to state their case personally. In this instance, access was a means of correcting an industry imbalance that the independents believed worked to their disadvantage.

This tradition of informal consultations had to be severely revised after the passage of the Administrative Procedures Act and the Open Meetings Act in the mid 1970s.[4] Texas officials are now forbidden to "communicate, directly or indirectly, in connection with any issue of fact or law," except in a public meeting. These laws have caused considerable distress on the Commission and within the independent portion of the industry. Commissioners are not certain that they can legally consult even with their staff; they can certainly not personally discuss field rules with members of the industry.

The recent growth of independent oil and gas consultants has kept the independent producers from panicking over the loss of their access to Commissioners, but they are still greatly displeased. Pressure from the industry and from state agencies may soon persuade the state legislature to modify these laws.

PERSONAL INFLUENCE AND POLICY MAKING

There is another kind of personal influence, however, that is of more interest to students of the policy process. This is the network of friendships with individuals in the industry that have traditionally been maintained by Commissioners.

Students of democratic government have often been made uneasy by social ties between regulators and the regulated. The occupants of regulatory positions should, in theory, be detached and impartial. They can hardly be expected to objectively arbitrate the conflicts between industries and the larger society if they have emotional ties with members of those industries. A common assumption for those who observe Washington regulatory agencies, therefore, has always been that friendships between government regulators and people in the industries they regulate are undesirable.

Observers of regulatory bodies are concerned that the people who serve on them will become coopted—that is, that they will come to endorse the industry's view of reality and thus lose sight of the greater public's interest. Their view is that a regulator who fraternizes with the regulated is almost certain to be coopted. They have consequently warned against social relations between people on regulatory agencies and people in the regulated industries.[5]

In the federal government, the principle of nonfraternization has never been assiduously observed. But, on the Railroad Commission, it is not even acknowledged. The friendships of many Commissioners with individuals in the industry have been open and unabashed. Many Commissioners have maintained casual friendships with certain oil people and have counted others among their closest companions. Nor have members of the industry been ashamed to proclaim their affection for various Commissioners; when Olin Culberson died in 1961, for example, several of his pallbearers were independent oilmen.[6]

But the fact that Texans have not seemed to worry about the inherent conflict of interest in personal ties between regulators and the regulated does not mean that it is not a problem. Commissioners, naturally, insist that they make all their decisions on the basis of enduring principles and objective evidence, and perhaps they do. Perhaps they forget who their friends are when they sit down to make policy. There is nevertheless something unseemly about the constant association and eternal good humor between Commissioners and the industry.

It would be a mistake, however, to suppose that whatever influence exists in these friendships flows only in one direction. The assumption of most critics of American regulatory agencies is that, through personal contact, the industry influences the behavior of the regulators. This is perhaps true, but it is an incomplete view, because it ignores the fact that, in friendship, influence is mutual. Regulators are capable of influencing through personal contacts as well as being influenced.

In the case of the Railroad Commission, it seems that some Commissioners have employed personal relationships to extend their own authority. That is, they used personal influence to make an informal impact on the industry over and above the formal impact they were able to make in their official capacities. Without stretching the facts too far, we could say that they sometimes used the industry to serve their own purposes as public servants.

The best example of a Commissioner who influenced the direction of the industry through personal contacts is Colonel Ernest Thompson. An examination of his method of governing the industry will illustrate the symbiotic nature of influence between the industry and the Railroad Commission.

A CASE STUDY OF PERSONAL INFLUENCE: ERNEST THOMPSON

In his almost thirty-three years on the Railroad Commission, Ernest Thompson might have been expected to become well known among members of the industry. They could observe him in action at monthly statewide hearings or at the occasional field hearings that came before the Commissioners, and, if they had any propensity to attend trade association conventions, they would undoubtedly have seen him give one of his many speeches. It is not surprising to discover that, half a generation after his death, Thompson is remembered with clarity by nearly everyone in the Texas industry who was an adult when he was alive.

But the quality of this recollection *is* surprising. Thompson is not just remembered—he is revered. He is described as if he understood everything, never made a bad decision, and single-handedly ruled the world of petroleum with wisdom and decorum. Given the fact that Thompson's Commission was constantly embroiled in the many fractious disputes of the industry and that he was rarely reluctant to take sides in an industry argument, it is curious that he has left a legacy of such unanimous respect and affection.

The key to understanding Thompson's huge success at navigating the turbulent waters of industry politics lies in understanding his talent for interpersonal relations. Quite simply, the colonel was such an effective politician because he was able to make nearly everyone of any consequence in the industry his close friend. Fifteen years after his death, the Texas industry is still full of people who refer to "my great friend Ernest Thompson" with an emotion that they show for no other subject.

Some of the colonel's attractive qualities seem to have come to him naturally. He was universally acknowledged to have a brilliant intellect. Although a lawyer and a military man by training, he grasped difficult economic and engineering concepts with ease and always understood which details were germane to general points. Moreover, his fearlessness and vigor were legendary in the industry. But there are many bright, energetic people who alienate others instead of winning their allegiance. Simply enumerating Thompson's strengths of character is not enough to account for the reverence in which his memory is held.

It seems, rather, that the colonel was so effective in making friends because he worked at it with skill and determination. He systematically cultivated relationships with members of the industry. When a newly assigned crude-oil buyer for a major company first arrived in Austin, for example, the man was certain to be asked over to Thompson's house for a drink. There he would be regaled with stories, told a few inside tidbits about the Commission, and generally made to feel like a treasured companion. If he was a drinker, he and the colonel might stay up late together, swapping jokes or war tales, downing their favorite alcoholic beverages, and becoming trusted intimates. At the end of one of these sessions, the colonel had a new friend, and the industry had a new admirer of Ernest Thompson.

Thompson would not allow these friendships to lapse. His social life consisted of keeping up his contacts within the industry. At the end of every monthly statewide hearing, he would mingle with the crowd of major company representatives, soothing with generous applications of charm any egos that might have been bruised by his caustic questions. His vacations might be spent traveling in the company of one or another of the Texas independents. He attended industry conventions and was a guest at industry parties. As a result of this constant pursuit of good-fellowship, the domestic world of oil revolved around Thompson. He came close to being its social chairman as well as its chief rule giver.

But the two roles complemented each other. Thompson was able to enforce political peace on a quarrelsome industry because the one common denominator of the battling factions was their respect for him. In particular: it might seem peculiar that the major companies never launched a political offensive against the Railroad

Commission because of its sustained partiality to the independent wing of the industry. By the 1940s, it was patently obvious that in every significant policy—well spacing, allocation, field allowables—the Commissioners were determined to give the independents considerably more than an even break. Every year, Commission regulations cost the majors millions of dollars. Yet the majors never tried to wrest political control of the Commission from the independents. Given their tremendous resources, the majors might have been expected to sponsor electoral challenges to Commissioners, or to try to bring in the federal government, or in some other way to undermine the Commission.

But to attack the Railroad Commission was to attack Ernest Thompson. The colonel was the cherished friend of many of the top executives of the major companies, as well as of most of their operatives in Austin. So great was their affection for him that it never occurred to them to try a frontal assault on his policies. Instead, they contented themselves with challenging the Commission's rules in the more restrained and impersonal arena of the courtroom. Ernest Thompson's private influence kept the Commission tranquil and supreme amid the constant public guerrilla warfare of the domestic petroleum industry. It is a fine point, therefore, whether the industry coopted Thompson or Thompson coopted the industry.

CREATING CONSENSUS

Personal influence has always been central to Railroad Commission politics, but organized influence is important, also. Members of the Texas industry share an interest in being able to concentrate their resources to meet or prevent general widespread opposition to Commission policies. From the 1930s to 1972, there was in Texas an undercurrent of dissatisfaction with prorationing, and since 1973 a proconsumer sentiment has consistently viewed the Commission with suspicion. Yet, for five decades, every Commissioner has been favorably disposed toward the industry. No doubt this is partly because Commissioners have traditionally believed that "what's good for the petroleum industry is good for Texas." But it is also due to the demonstrated capacity of people in the industry to agree among themselves to support or oppose certain candidates. As will be explained in chapter 8, the united voice of the industry is a powerful force in Commission elections. It is therefore of value to explore the manner in which the dispersed and heterogeneous Texas industry forms a political consensus so that its energies are not dissipated in opposing itself at election time.

People in the industry follow no single voice when making up their minds about politics. There is no journal whose editorials are taken as authority, no organization whose official pronouncements are followed by everyone, and certainly no one person who directs everyone else's opinions. Indeed, industry members, especially the independents, frequently possess a self-reliance of judgment that borders on temperamental anarchy. There is no institution or personality that can be said to give formal leadership to opinion in the Texas petroleum industry.

Instead, the process of building consensus is an informal one. It results from the personal contacts of individuals as they encounter one another in a variety of settings. Of these settings, the most important are probably the petroleum clubs that are found in every large Texas city. These clubs furnish an arena within which members of the industry can eat lunch and enjoy fellowship in a congenial atmosphere and, incidentally, discuss public affairs.

The clubs are found in the cities' downtown business sections, close to the main offices of many oil companies. The Houston Petroleum Club, for example, is located on the top floor of the Exxon Building, the Dallas club on top of the First National Bank Building. During the lunch hour, these clubs are packed, and, as they serve superior food, they are popular for dinner as well.

At any given noon hour, dozens of oil people gather in the petroleum clubs to eat and chat. There need be no purposeful concentration on politics during these conversations; there does not have to be. Public affairs is as natural a topic of conversation as is business or sports or the weather. Similarly, many members stop in after work for a drink before going home, which offers another opportunity for conversation.

Besides serving as forums for informal discussion, the petroleum clubs hold social functions—parties and dances which provide an opportunity for members of the industry to get to know one another while they have fun. The most important social event, however, is somewhat more cosmopolitan. Every February the Dallas Petroleum Club hosts a formal ball, to which not only the most influential of the oil fraternity but important bankers, industrialists, sheiks, and politicians from within and without Texas are invited. At this ball the elite of Texas meet the elite from the rest of the world, as society, business, and politics meld together for an evening.

Despite the importance of this one February ball, the petroleum clubs are useful primarily for helping form a group mind within each city. For the more difficult job of creating a statewide industry perspective, other institutions are necessary. This task of maintaining contacts between members of the industry in different sections of the state is largely performed by the trade associations.

As with other industries, petroleum supports a wide range of these groups, from the small and specialized—like the West Central Texas Oil and Gas Association in Abilene—to the major umbrella organization—the American Petroleum Institute in Washington. Through their interactions at the meetings sponsored by these associations, individuals in the industry are helped to make up their minds about politics.

Incomparably the organizations most vital to the process of consensus building in the Lone Star State are the Mid-Continent Oil and Gas Association (membership approximately 3,200) and the Texas Independent Producers and Royalty Owners Association (TIPRO—membership about 4,200).

Political scientists have studied the formal activities of business trade associations at some length. These groups lobby regulatory agencies or legislatures, compile technical information for both their members and the government, engage in public relations campaigns, and so on.[7] Mid-Continent and TIPRO fall easily into the common pattern. Mid-Continent speaks for the Texas industry as a whole, TIPRO only for its independents. But the great importance of these two associations to Texas politics lies not so much in their official activities as in the forum they provide for interpersonal contact among industry members.

Each organization holds a yearly convention. Mid-Continent's is usually in October, TIPRO's in late spring. The conventions last two to three days and are full of the expected activities—speeches, committee meetings, banquets, awards ceremonies, and the like. Many of the official functions—for example, the catered meals—provide an opportunity for personal interaction. So also does the dance that is invariably scheduled. To these planned activities are always added the spontaneous meetings that result when friends are thrown together. There are all-night poker games, conversations in elevators, and shopping expeditions.

Throughout all these encounters, the members talk shop. Much of the conversation concerns business only. But some of it is inevitably political in nature: are we going to be able to beat that windfall profits tax in Congress? Can you figure out those new Federal Energy Regulation Commission rules? Who do you like for Railroad

Commissioner in the Democratic primary? The discussions go on throughout the convention. When they are finished, the industry has moved a step toward making up its collective mind about politics.

Although Mid-Continent and TIPRO are the most inclusive, and consequently the most important, cogs in the statewide machinery of informal discussion, smaller groups continue the process. There are, as mentioned, local trade associations. There are also professional organizations. The Texas chapter of the American Institute of Mining Engineers holds an annual convention, at which petroleum engineers have a special meeting. Additionally, there are so many engineers in Houston that they have their own suborganization and meet twice a month to have lunch, hear a speaker, and maintain friendships. Petroleum lawyers have a section of the annual state American Bar Association convention devoted to them. And so it goes. Through its network of overlapping organizations, the Texas industry is engaged in an almost constant internal dialogue concerning business, personal relations, and public affairs.

But industry institutions do not provide the only opportunities for personal interaction and deliberation. The Railroad Commission does its part, also. Every month, all the petroleum buyers and many sellers come together at the statewide hearing where purchasers nominate the quantity of oil they wish to buy in the coming month. In the old days of prorationing, these gatherings were an important event in the world of petroleum. But, when allowables went to 100 percent in 1972, it seemed that they had become superfluous, for purchasers were now willing to buy all the crude that the state could produce. The relatively inexperienced Commissioners of the late 1970s saw a chance to save themselves some time, and in June 1978 they announced their intention to institute quarterly, as opposed to monthly, statewide hearings.[8]

They were soon persuaded to revise their intention. Even with nominations of no further consequence, the statewide hearings were nevertheless vital to crude-oil purchasers. The monthly trip to Austin gave major company representatives a chance to meet, swap oil stocks, and talk shop. A voluntary meeting would smell of collusion and provide industry critics with an excuse to accuse the companies of antitrust violations. A gathering in obedience to a Commission summons, however, relieves the companies of responsibility, thus protecting them. When this was explained to the Commissioners, they quickly dropped their plan to abandon the monthly hearings.

Finally, every spring the Commission provides the major companies and larger independents with a forum where they can air their grievances and publicly discuss the future of the petroleum business. The Commission invites the top executives of the big companies to come to a state-of-the-industry hearing. One speaker from each corporation is asked to give a fifteen- to twenty-minute assessment of that company's present economic situation and future prospects and to comment on current trends in public policies affecting the industry.

There is some genuine information communicated at these meetings, but in general they are mostly an opportunity to "preach to the choir." That is, the executives explain the necessity of decontrol of petroleum prices, exhort the nation to avoid windfall profits taxes, criticize the federal Department of Energy, and give one another and the Commissioners all sorts of other superfluous advice. These speeches, therefore, are largely for the emotional gratification of the participants. The annual state-of-the-industry gathering has, however, another, less publicized purpose.

All the executives usually stay at the same hotel, and the night before the hearing it inevitably witnesses a large party at which both business and politics are mixed with the socializing. Serious issues are discussed, courses of action proffered, positions clarified, and pulses figuratively taken. As with the monthly statewide hearings, this annual gathering allows many important people in the petroleum industry to counsel together without making themselves vulnerable to antitrust suits. Once again, the Railroad Commission holds an umbrella of legitimacy over an industry gathering that is directly relevant to the politics of the Commission.

In addition to the interactions that are in some sense structured by trade associations and Commission hearings, there is another, even more informal kind of organization that is of importance to the Railroad Commission. This is what might be termed influence by activity.

Students of interest groups have pointed out that the burden of responsibility in any group is likely to be carried by a relatively small proportion of its membership. It is this active minority, as opposed to the relatively apathetic majority, that determines the group's policies, raises and spends money, speaks on behalf of the group, and, in general, makes the group's decisions.[9]

The active minority is also of great importance to industry political participation, but in a way less structured than the one described by students of interest groups. Group theorists emphasize that the minority of any group uses that group as an institution. The resources of the group are used by the active minority to advance what they conceive to be the group's goals.

But, in the petroleum industry, many individuals are so wealthy and well connected that they do not need to work through an organizational structure. If there is a candidate to be supported, for example, a single, identifiable congregation of individuals in the larger Texas cities is likely to organize spontaneously to raise money, generate publicity, make sure that the candidate meets the right people, and so on. These individuals are invariably the wealthier independents, and their activities exert a decisive effect on Railroad Commission politics. The part they play in campaigns will be explored in chapter 8.

Despite the fact that the Texas petroleum industry is composed of thousands of firms and tens of thousands of individuals spread over a relatively large area, therefore, it is often able to create an informal consensus on matters of public policy. When there is a clash of interests, of course, as in the import controversy of two decades ago, no industry agreement on a course of action pertaining to that problem is possible. As will be shown in the next chapter, however, the industry often manages to attain near unanimity on such important topics as which candidate to support in Railroad Commission campaigns.

The petroleum industry has great resources, but that would be of little consequence if the various segments of the industry used their resources against one another. The informal network of communication and friendship within the industry serves to dampen conflict, thus allowing industry effort to be focused. It will become clear in the next chapter that this focused effort has always been irresistible in Commission campaigns.

But there is another arena in which the influence of the industry is more opaque: gubernatorial appointments. The strengths and weaknesses of industry influence can be better understood by examining the pattern of decision making that produces new members to the Commission when a seat becomes vacant between elections.

GUBERNATORIAL APPOINTMENTS

For a variety of reasons, most Railroad Commissioners have first attained the position through appointment rather than election. One Commissioner has died while serving; others have resigned to seek higher office, take a job in private industry, avoid a conflict of interest, or the like. Midterm resignations are so common that, between 1930 and 1980, only three people initially attained a Commission seat by winning an election: Jerry Sadler in 1939, Olin Culberson in 1941, and Jon Newton in 1977. In the same period, governors have appointed nine men to the Commission. Having a say in gubernatorial appointments would clearly be a very substantial influence over the politics of the Commission. If we are to understand the relationship between the industry and the Commissioners, therefore, we must have some notion of the extent to which the industry influences Texas governors when they choose new Commissioners.

Because the power to make appointments to state agencies obviously involves the power to affect public policy, scholars of state politics have usually ranked it as one of the most important of gubernatorial tasks.[10] Yet, although there are studies of governors' constitutional and statutory authority, no political scientist has examined the *process of choice* engaged in by governors when making an appointment to a state agency. The present inquiry must therefore be very basic, since it has no previous research on which to build.

There are several alternative patterns that might apply to the relationship between governors, potential appointees, and the industry. It might be expected that one of five models would apply to the normal process of appointment:

1. The "civics course" model. It is possible that governors merely appoint the best-qualified person for the job, taking only that person's expertise and integrity into consideration.

2. The industry-dominant model. Alternatively, governors may, in effect, turn over the appointment process to the industry. Governors, without exercising much personal discretion, may ask individuals within the industry to provide the industry's consensus choice.

3. The governor's list model. Or, governors may circulate a list of names and appoint the person from this list most favored by the industry. Something like this frequently occurs in appointments to the United States Supreme Court. When there is a vacancy on the court, the president often asks the American Bar Association to offer its comments on a list of potential nominees. In the case of the Railroad Commission, this process would be less formal, but the principle is the same.

4. The industry's list model. As a fourth possibility, industry representatives offer a list of acceptable candidates to the governors, who then pick the one who most appeals to them.

5. The industry's veto model. Finally, it is possible that the in-

dustry has little influence over the governors' choice but may stop an appointment it finds repugnant. Governors exercise great freedom in their decision, but, if they should choose a candidate whom the industry finds offensive, that candidate could be prevented from taking office.

Perhaps surprisingly, all available evidence points to the industry's veto model as the one most applicable to Railroad Commission appointments. In the last twenty years, governors have never appointed a person to the Railroad Commission who was generally hostile to the industry, but neither have they picked individuals who were dictated by, or even suggested by, the industry.

When there is a midterm vacancy on the Commission, of course, governors receive many recommendations from acquaintances in the industry concerning who would be the best person for the job. Lists of individuals who are commonly thought to be well qualified for the seat are discussed by people who take an interest in the Commission. Such lists often find their way into the newspapers, where they turn private discussion into public speculation.[11] But, in the last twenty years, governors have never picked one of these front-running industry candidates—quite the contrary. They almost invariably choose someone they trust personally, someone who has little or no previous experience with the oil and gas business. Such men have always been more or less favorably disposed toward the industry but are unknown within its confines. Their appointment normally comes as a complete surprise even to those industry members who know the governor personally.

For example, when Olin Culberson died in 1961, Governor Price Daniel appointed Ben Ramsey. In over a decade as Texas' lieutenant governor, Ramsey had proven himself a loyal conservative Democrat, but he had no particular experience with oil and gas matters.[12] When William Murray resigned amidst charges of conflict of interest in 1963, Governor John Connally needed an appointee of unquestioned integrity. He picked Jim Langdon, a former classmate at the University of Texas School of Law and, at the time, chief justice of the Circuit Court of Civil Appeals. When resignations created Commission vacancies in 1973 and 1977, Governor Dolph Briscoe appointed two members of his personal staff—Mack Wallace, a former county district attorney, and John Poerner, a former state legislator and a land surveyor by profession. The elevation of all four of these men was rather startling to the industry, for none was known inside petroleum circles. The industry quickly came to respect and support all four, but that does not mean that it had a hand in their selection.

Despite the fact that governors are quite independent of the industry in making appointments, however, all observers are agreed that the industry could veto their choice. In the first place, no recent Texas governor would have made an appointment to the Commission over the vocal opposition of an important segment of the industry. All governors since the eve of World War Two have been friendly to oil, so they would not purposely antagonize the industry. But, even if a proconsumer, antipetroleum governor is elected in the near future (an unlikely event), industry members are confident that they could still stop a hostile appointment. Under the Texas Constitution, all gubernatorial appointments must be confirmed by the state senate. Industry members express no doubts that their clout in the Senate is more than sufficient to prevent any antioil candidates from being confirmed.

While the industry does not dominate the process of choosing Railroad Commissioners between elections, therefore, it exercises a real, if negative, influence over the process of choosing them.

CONCLUSION

The fact that there is mutual influence between Railroad Commissioners and the people in the industry they regulate does not mean that, in their day-to-day decisions, Commissioners do not make defensible choices, based on sound principles and evidence. The regulation of petroleum is an exacting activity, requiring intelligence and strength of character. There is no reason to suspect that individual decisions by Commissioners are in any way the product of pressure or favoritism. It is instead the *tendencies* of policy that are formed in the interactions between the industry and the Commissioners. The exchanges of influence have determined not specific decisions but the broad directions and long-run purposes of the Commission.

The application of influence in the politics of the Railroad Commission is quite complex. It consists partly of formal organization, partly of the informal by-product of formal organization, and partly of individual initiative. The petroleum industry influences Commissioners, but Commissioners in turn influence the industry, both in their official capacities and through personal relations.

Part of the effort made by the industry to influence the Commission consists of an attempt to dominate the channels of selection to the agency. As this chapter has demonstrated, the industry exerts influence over one of those channels—gubernatorial appointments—but its power in that sphere is limited and after the fact. There is another and more important channel of selection, however: elections. If the industry could prevail over that avenue, it could almost certainly assure itself a friendly audience on the Commission. The industry does indeed dominate Railroad Commission elections. The next chapter will explore the role played by the petroleum industry in Commission campaigns and elections.

8. Campaigns and Elections

More people have been elected between sunset and sunrise than ever were elected between sunrise and sunset.—Will Rogers

Regardless of how much the governor has to say about who becomes a Railroad Commissioner, all those who serve on the agency must survive the test of electability. Commissioners hold office for staggered six-year terms. Individuals appointed because of a midterm resignation must face the voters within two years after getting the job; if they survive, they must again defend the seat at the normal end of their predecessors' term, which may be another two, four, or six years. At least one of the three Commission seats is consequently vulnerable to electoral capture every two years. Whatever the other political influences on the Commission, the controlling reality of Commission life must therefore be the need to survive challenges at the polls.

In this regard, the Railroad Commission is unique. The Commission is not the only state regulatory agency whose personnel must run for election. But it is the only body of national importance in that position. No other state regulatory body, whether appointive or elective, approaches the national influences of the Commission; no federal regulatory agency is elective. The Railroad Commission is consequently the only nationally relevant regulatory body in the country whose policies are directly under the influence of voters.

PRIVATE INTERESTS AND PUBLIC AUTHORITY

As explained in chapter 2, it is no accident that the Commission is an elective body. The guiding force behind its establishment in 1891, Governor Jim Hogg, believed that an elected agency would be subject to the corruptive power of railroad campaign resources. He thought that the only way to keep it identified with the public interests was for its personnel to be dependent not on electoral campaigns but on himself. The original Commission was consequently appointive. Political circumstances and a blunder by Hogg, however, combined to force a change, and the Commission became elective three years after it was established. In the following years, the suspicion that an elective regulatory agency was too vulnerable to capture by private interests was occasionally revived. In the 1930s, for example, there was a strong movement in Austin to transfer the Railroad Commission's authority to regulate oil and gas to a new, appointive agency. Such proposals have always failed, and after ninety years of operation the Commission is still elective.

But the issue raised by Hogg is at least as important today as it was in the 1890s and 1930s. The industry has changed, but the problem remains: can a private interest dominate a public authority because of its superior campaign resources? This chapter will provide information that is directly relevant to this question.

PERSONAL ORGANIZATION

In order to attain or retain a seat on the Commission, individuals must run a campaign without effective party backing in a huge state. In such an environment, the candidate with the greatest resources begins with an advantage. As will be explained later in this chapter, the resource relied upon by all recent winning candidates has been money. Successful Commission candidates, however, have also occasionally enjoyed access to another kind of campaign resource: personal organization. Some Commissioners have been able to mobilize a network of volunteer campaign workers to ring doorbells, hold fund-raising parties, distribute literature, and, in general, approximate a political machine. The impact of these personal organizations is not measurable, but it must have been substantial.

For example, Olin Culberson was state secretary of the Texas Firemen's and Fire Marshals' Association. In the decades during which he was in politics, there was a voluntary fire brigade in most small towns in the state. These associations formed the backbone of his campaign organization in the rural areas in the 1940s and 1950s.

Having been a conspicuously successful mayor of Amarillo, Ernest Thompson was popular with many other mayors in the state, who used whatever influence they possessed in their own cities on his behalf during campaigns.

As the only petroleum engineer on the Railroad Commission, William Murray was frequently asked to speak at gatherings of those in that profession. He rarely refused an invitation. He was a long-time scoutmaster and a member of the national executive board of that organization. Both petroleum engineers (of whom there are thousands in Texas) and Boy Scouts could be counted on to participate in his reelection efforts.

Before they were appointed to the Commission, Ben Ramsey and Byron Tunnell had both been active in Texas politics for years. They had many contacts around the state and a good relationship with many interest groups. When they had to defend their seats on the Commission, the personal organizations that they had developed adjusted easily to running their campaigns for the new office.

The ability to forge a statewide campaign network *before* one runs for, or defends, a seat on the Railroad Commission thus appears to be of considerable importance in making a candidate electable. It is probable, however, that this is true in any political race in any state. The conclusion that the candidate with the best organization frequently wins will not come as startling news to anyone with more than a passing knowledge of American politics. The real problem involves determining in what ways campaigns for the Railroad Commission are unique, not in what ways they are familiar. In order to do this, it is necessary to shift the focus from the candidate to the constituency and ask: how do the structure and activity of the petroleum industry affect Commission elections? Is the industry just another interest group, or does it dominate Commission campaigns? If so, how? What can we say about democratic regulation of oil and gas in Texas?

THE INDUSTRY AND ELECTIONS

As explained in chapter 7, the petroleum industry exerts only partial and negative control over gubernatorial appointments. But that is of secondary importance to its potential influence over elections. The rest of this chapter will explore industry efforts to dominate Commission elections. Possible bloc voting as well as campaign contributions will be examined. The conclusion will be that bloc voting is not important but that contributions are important to industry control over election outcomes. Then the elections of 1976 and 1980 will be discussed to illustrate the industry process of gathering and using contributions to influence campaigns. Finally, in chapter 9, the question posed by Governor Hogg will again be addressed.

BLOC VOTING

As a state with a southern history, Texas has had consistently low voter-turnout rates. Since 1960, about two-fifths of its adult citizens can be expected to go to the polls in presidential years, perhaps one-third in off years.[1] The oil and gas industry is extraordinarily important to Texas, both in terms of the amount of money it pumps into the local economy and in terms of the number of people it employs directly or indirectly.[2] Given the low turnout, if the industry could reach a consensus on a candidate for the Commission it might very well be able to determine electoral outcomes by bloc voting.

We will look at a series of election results for the Railroad Commission and try to determine whether this hypothesis represents reality. Only Democratic primaries will be relevant to this effort, for no Republican has ever won a seat on the Commission.

Texas has 254 counties. Although the petroleum industry is represented in every part of the state, it is not equally important to each county. Some 40 counties have never produced petroleum, and the importance of the industry in the remaining counties varies widely. In order to estimate the effect of the industry on voting, we must devise some measure of the importance of petroleum to the population of each county. Then, we can see whether the vote for candidates friendly to the industry varies with the salience of the industry in each county. If support for the industry candidate is greater in counties where oil and gas are of more importance to the local economy, then it stands to reason that bloc voting is a reality. If support does not vary with the importance of the industry, then bloc voting is not a defensible interpretation.

Ideally, to measure the importance of the industry to the population of each county, we should know the proportion of a county's work force that is employed, directly or indirectly, by the petroleum industry. Such figures are not available, however, so a less direct measure will have to be found. The most logical indirect method of assessing the economic impact of petroleum on a county is to measure the amount of oil production for each person in that county. Because of the historic Railroad Commission policy of encouraging close drilling, the more oil production in a county, the more wells there are likely to be. Other things being equal, the more wells there are, the more people are needed to tend them, the more landowners receive royalty checks, and so on. The more oil produced in a given county per person, therefore, the more important the petroleum industry is likely to be to the economy of that county.

The Mid-Continent Oil and Gas Association compiles yearly summations of the number of barrels of oil produced in each Texas county. Additionally, it is possible to estimate the total population in each county in each year, using the decennial federal census figures and some simple arithmetic. When the estimated number of inhabitants in each election year is divided by the number of barrels of oil produced in that year, the result is per capita oil production—that is, the number of barrels of oil produced in each county per person in the given year. For this chapter, it will be assumed that a county's per capita oil production is an accurate indicator of the industry's importance there.

There must also be some objective way of determining the industry candidate in each election. For this purpose, it will be assumed that any candidate who received 70 percent or more of the total campaign contributions of the petroleum industry is the industry candidate. Normally, this favored candidate receives much more than the 70 percent minimum; 97 percent is not uncommon. In practice, this candidate won every primary and every general election in the 1960s and 1970s.

One of the most sophisticated mathematical tools for analyzing voting records is the stepwise multiple regression. In this technique, independent variables are fed automatically into a regression equation by the computer, in the order in which they can statistically explain the variance in the dependent variable. For each independent variable, there is a Beta slope, showing its effect on the dependent variable; a cumulative r-squared, indicating the amount of variance explained, after the variance explained by the preceding variables is subtracted; and a measure of statistical significance. The equation as a whole also has a significance level. This technique is useful in examining voting behavior in elections in which there are many possible influences on the results.

For this study, the winner's percentage of the vote in all eight contested elections, from 1962 to 1978, was the dependent variable in the multiple regressions. Independent variables were the proportion of the population in each county that was urban, black, or chicano, according to the 1970 census; the proportion of the voting-age population that was registered to vote in each election year; and the median income and per capita oil production of each county in each election year.[3] If there is any relationship, either independent or interactive, between any of these variables and the vote for the industry candidate, it will show up in these equations.

As table 1 illustrates, there are occasional relationships be-

Table 1. Democratic Primaries for the Railroad Commission, 1962–1978, Stepwise Multiple Regressions

Independent Variable	*B*	*Cumulative* r^2	*Significance*	*Significance of Entire Equation*
Dependent Variable: Winning %, 1962				
MEDINC70	−.26	.13	$p < .001$	$p < .001$
PRCPOP62	−.74	.17	$p < .001$	
PCBLK70	.11	.178	not sig.	
PCCHI70	.17	.18	not sig.	
PCURB70	.30	.19	not sig.	
REGVOT62	−.15	.192	not sig.	
Dependent Variable: Winning %, 1964				
PCBLK70	−.14	.02	$p < .01$	$p < .01$
PCURB70	.33	.03	$p < .01$	
PCCHI70	.25	.04	not sig.	
REGVOT64	.19	.045	not sig.	
PRCPOP64	.22	.049	not sig.	
MEDINC70	.47	.05	not sig.	
Dependent Variable: Winning %, 1966				
PCURB70	.61	.008	not sig.	not sig.
REGVOT66	−.49	.019	not sig.	
PCBLK70	.11	.03	not sig.	
PCCHI70	−.20	.035	not sig.	
PRCPOP66	−.33	.039	not sig.	
MEDINC70	−.42	.04	not sig.	
Dependent Variable: Winning %, 1970				
PCURB70	.32	.01	$p < .01$	not sig.
PCBLK70	.62	.02	not sig.	
MEDINC70	−.27	.025	not sig.	
PRCPOP70	−.13	.027	not sig.	
PCCHI70	.30	.0273	not sig.	

Independent Variable	*B*	*Cumulative* r^2	*Significance*	*Significance of Entire Equation*
		Dependent Variable: Winning %, 1972		
PCBLK70	−.10	.01	$p < .01$	not sig.
PCCHI70	−.15	.02	not sig.	
PRCPOP72	.12	.025	not sig.	
MEDINC70	.26	.026	not sig.	
PCURB70	−.13	.027	not sig.	
REGVOT72	.44	.028	not sig.	
		Dependent Variable: Winning %, 1974		
PCCHI70	−.64	.05	$p < .001$	$p < .001$
MEDINC70	−.11	.07	$p < .01$	
PCBLK70	.15	.08	$p < .01$	
PRCPOP74	.42	.088	not sig.	
REGVOT74	−.31	.09	not sig.	
		Dependent Variable: Winning %, 1976		
PCURB70	.15	.18	$p < .001$	$p < .001$
PCBLK70	−.26	.23	$p < .001$	
MEDINC70	−.71	.235	$p < .01$	
PRCPOP76	−.20	.238	not sig.	
PCCHI70	−.10	.24	not sig.	
REGVOT76	.71	.242	not sig.	
		Dependent Variable: Winning %, 1978		
PCURB70	.10	.12	$p < .001$	$p < .001$
PCBLK70	−.24	.17	$p < .001$	
MEDINC70	.79	.172	not sig.	
PRCPOP78	−.26	.176	not sig.	
PCCHI70	.32	.1771	not sig.	
REGVOT78	.25	.1772	not sig.	

Key

Winning %	Proportion of the total vote going to the winner	PCCHI	Percentage chicano
MEDINC	Median income	PCURB	Percentage urban
PCBLK	Percentage black	PRCPOP	Per capita oil production
		REGVOT	Registered voters

tween these variables and the winner's percentage of the vote, but they are neither large enough nor consistent enough to warrant close attention. In 1962 there is a moderate negative association between median income and winning percentage, but it disappears thereafter. In the 1976 and 1978 runoff elections, urban counties tend to support the winner more than rural counties. As the losing candidate in those races was a veteran East Texas politician with an oft demonstrated appeal for rural voters, it seems clear that his greater support in the countryside is not an indication of a trend that will outlast his candidacy.

Other relationships crop up occasionally in individual elections in table 1, but none is of enough strength to be interesting, and none carries over from one campaign to another. In particular, per capita oil production shows a significant relationship with voting only in 1962, and that year the association is negative. In other words, county-level analysis strongly suggests that people associated with the petroleum industry do not bloc vote for industry candidates for the Railroad Commission in Democratic primaries. It appears that the industry candidate won every primary in those years because he received the votes of the majority of citizens in general. If the industry dominates elections, it must be because it is able to induce Texans who are not connected with petroleum production to support its chosen candidates.

CAMPAIGN CONTRIBUTIONS

There must be some way that the industry ensures widespread support among ordinary citizens for its favored candidates. The most plausible explanation is that the industry candidates are able to put on a much more effective advertising campaign than their rivals.

In a general election, in which candidates from two or more parties compete, most of the voters are helped to decide on their choice by knowing the party labels of the candidates. Most voters consider themselves either Republicans or Democrats and use the party affiliations of the candidates as cues to guide their votes.[4] But in a primary such cues do not exist, for of course all candidates belong to the same party. In such a situation, voters are inclined to pick candidates who have achieved name recognition and a favorable personal image.[5]

The Railroad Commission primary race is almost an ideal type

Table 2. Campaign Contributions to Industry and Nonindustry Candidates, Railroad Commission, 1962–1978 Democratic Primaries

Year	*Total Contributions*	*Total Contributions to Industry Candidate*	*Percent of Total to Industry Candidate*	*Number of Nonindustry Candidates*
1978[a,b]	$406,969.18	$393,764.18	97	3
1978	126,350.00	126,350.00	100	0
1976[a]	439,353.58	341,118.59	78	6
1974[b]	61,499.00	61,434.00	99.9	1
1974	50,720.00	50,720.00	100	0
1972	48,957.35	47,349.35	97	2
1970	40,125.00	40,125.00	100	1
1968	3,550.00	3,550.00	100	0
1966	114,127.49	109,754.50	96	2
1964	5,950.00	5,950.00	100	0
1964[b]	164,996.47	141,535.22	86	1
1962[b]	90,961.00	90,961.00	100	0

[a]Includes runoff.
[b]Unexpired term.

of what voting-behavior specialists call a low-stimulus campaign. That is, it is not of much interest to most voters and receives relatively little attention from the news media. Other races, such as those for the presidency, Congress, and the governorship, attract most of the publicity. When voters arrive at the polling booths on election day, they will probably have received some information about high-stimulus races, but they may have heard little or nothing through the news media about such low-profile races as those for the Railroad Commission.

In a low-stimulus campaign, any candidate whose name becomes widely and favorably known has a great advantage. Manufacturing this sort of good image is the function of advertising. Advertising, especially in the electronic media, is expensive. Other things being equal, a candidate with an edge in campaign financing has a superior opportunity to establish name recognition through ad-

Table 3. Majorities for Industry Candidate, 1976 Democratic Primary Runoff, Selected Precincts

City	*County-wide*	*Wealthy White*	*Working-Class White*	*Black*	*Chicano*
Austin					
% of vote to industry candidate	74	83	63	60	70
Dallas					
% of vote to industry candidate	70	88	58	54	67
Houston					
% of vote to industry candidate	70	89	57	66	51
San Antonio					
% of vote to industry candidate	82	94		81	77

SOURCE: County courthouses for the relevant cities.

vertising and, consequently, is likely to win in a low-stimulus campaign.

If the petroleum industry candidates are found to receive a preponderance of campaign contributions, it would be relatively safe to assert that this is the source of their general voter support. They win because they have been well marketed, not because the industry bloc votes for them.

Table 2 establishes that, in the 1960s and 1970s, the industry candidate always enjoyed an immense financial advantage over his rivals.[6] Even in 1976, when six nonindustry candidates ran, they managed to attract less than 23 percent of the total contributions between them (more on this shortly). More commonly, the industry candidate received close to twenty times the total of his opponent or

Table 4. Majorities for Industry Candidate, 1978 Democratic Primary Runoff, Selected Precincts

City	*County-wide*	*Wealthy White*	*Working-Class White*	*Black*	*Chicano*
Austin					
% of vote to industry candidate	88	88	73	78	69
Dallas					
% of vote to industry candidate	76	93	60	63	65
Houston					
% of vote to industry candidate	70	92	57	67	67
San Antonio					
% of vote to industry candidate	73	91		74	63

SOURCE: County courthouses for the relevant cities.

opponents. It seems likely that this huge advantage in finances allowed the industry candidate to purchase the attention of ordinary people. Normally, he won because he was slightly known and his opponents were unknown.

This pattern of across-the-board support for industry candidates is illustrated in tables 3 and 4, which show precinct returns in the 1976 runoff between Jon Newton and Jerry Sadler and the 1978 runoff between John Poerner and Sadler. These precincts illustrate voting trends in wealthy white, working-class white, black, and chicano areas in Texas' capital and three largest cities.

Just as we would expect, Newton and Poerner, the industry candidates, walloped Sadler in wealthy white districts in all four cities. Less predictably, however, industry candidates won majorities in *every* section of town, and their margins of victory were often very large even in chicano and black precincts.

Superior finances thus have given industry candidates for the Railroad Commission opportunities to persuade ordinary voters to support their cause. Historically, the industry has not needed to bloc vote for its own candidates, because it has supplied them with overwhelming campaign resources.

CONTRIBUTIONS AND INDEPENDENTS

It is not just the origin of the favored candidate's electoral support that is illuminated by these figures. It has already been explained that the Railroad Commission has a long history of favoring small independent operators over major oil companies in its rulemaking. Since campaign contributions seem to be so vital to electoral success, it is reasonable to expect that independents buy this advantage over the majors by supplying the bulk of a candidate's war chest.

Table 5 enables us to confirm this suspicion. In this table, all campaign contributions to twelve winning candidates, from 1962 to 1978, have been examined. Contributors are required by law to reveal their names and hometowns. An effort was made to identify the profession of everyone who contributed five hundred dollars or more to each of the twelve candidates. Through the *Oil Directory of Texas*, lists of attorneys in telephone books, and many phone calls, better than 85 percent of the contributors were identified in eight out of the twelve cases.[7]

As table 5 illustrates, the hypothesis that independents dominate the contributors' list is strongly borne out by the figures. Normally, independents supply over 60 percent of the total, majors less than 5 percent. (The figures do not add up to 100 percent, because contributions from people in nonpetroleum businesses like trucking and railroads have been omitted.) Even if we assume that all lawyers who contributed money were really acting as conduits for the majors, the dominance of the independents is striking. The most reasonable explanation for the Commissioners' historical solicitude for the independents, therefore, is seen in the fact that the independents finance their campaigns.

Why don't the majors give more money? There seem to be several reasons. Majors have far-flung operations, both in other states and abroad, while independents are often so small that they produce only in Texas. Majors consequently do not rely on favorable treat-

Table 5. Sources of Contributions to Winning Railroad Commission Candidates, 1962–1978 Democratic Primaries

			Percent of Those Traceable		
Candidate/ Year	*Total Contributions of $500 and Over*	*Percent of $500 Contributions Traceable*	*Independent Oil and Gas Producers*	*Major Companies*	*Attorneys*
Poerner/78[a,b]	$288,576.63	92	59	2	7
Wallace/78	77,750.00	97	67	1	6
Newton/76[a]	285,012.78	83	73	0.1	5
Langdon/74[b]	41,965.00	86	63	0	18
Wallace/74	37,038.00	95	63	4	9
Tunnell/72	22,250.00	92	59	7	24
Ramsey/70	33,250.00	73	73	3	3
Langdon/68	1,500.00	67	50	0	0
Tunnell/66	22,400.00	89	89	0	0
Ramsey/64	4,000.00	88	43	0	43
Langdon/64[b]	97,570.00	39	80	0	16
Ramsey/62[b]	65,500.00	91	70	4	1

[a]Includes runoff.
[b]Unexpired term.

ment by the Commission, but many independents live or die by the policies of the Commissioners. Furthermore, while the majors are often billion-dollar companies, their wealth is corporate, not individual. Their top executives are well paid but not normally rich. Independent money, however, is often personal. The Hunts, Cullens, Richardsons, and thousands of lesser independents control fortunes that are theirs alone, with no corporate inhibitions. In addition, majors are forbidden by state and federal laws to use their corporate money directly for campaigns. Gulf tried to evade these restrictions, was detected, and received much bad publicity.[8]

So the structure of the industry and the legal context have combined to make independent money available for campaign contributions and major money unavailable. Although it is impossible to prove, it stands to reason that this has been a major factor in the friendliness with which the Commissioners have always treated the independents.

With the passage of campaign reform legislation in the mid 1970s, it became more common for corporations to indirectly contribute to candidates through Political Action Committees. Major company PAC money, included in the figures of table 5 for 1976 and 1978, caused no noticeable swelling in the proportion of contributions by the majors. Perhaps in the future, as the majors become more accustomed to using this money, their proportion of contributions to Railroad Commission candidates will increase.

A CASE STUDY OF INDUSTRY CAMPAIGN ACTIVITY: THE 1976 ELECTION

The portrait of the domination of Railroad Commission campaigns by the Texas petroleum industry is now relatively coherent. Thus, through a complicated network of personal contacts, people in the state industry have been able to form a consensus about whom to support in Commission races. Members of the industry, especially its independent wing, have made sure that this chosen candidate had a huge advantage in campaign financing. Through the advertising purchased with these funds, the industry candidate has been able to make sure that his name became known to many of the voters. He consequently won every primary between World War Two and 1978. He won the general election because Texans were overwhelmingly Democratic.

This explanation makes sense, but it contains a potentially fatal flaw. Because Commissioners have had a historical penchant for leaving office in the middle of their terms, the governor has appointed their successors. These new Commissioners have run for reelection as *incumbents*.

The presence of an incumbent candidate has two important effects on a campaign. First, it provides citizens with a voting cue. Research suggests that, given a list of unknown candidates, voters will often pick the incumbent.[9] If it is simply incumbency that attracts voters, we do not need superior finances to explain the victory of the industry candidate. Second, an incumbent has had a chance *before* the campaign to cultivate the industry. It is entirely possible that members of the industry decide individually to support an incumbent after observing him in office and do not go through the process of consensus building that was suggested in chapter 7. It might be possible to explain the normal attitudes of Railroad Commissioners by simply asserting that the governor appoints people who are favorable to the industry. Elections would be of secondary importance.

In our attempts to decide which interpretation is more correct, the 1976 and 1980 Democratic primaries will provide critical tests. The 1976 contest was the first in thirty-six years in which there was no incumbent running. The 1980 race was the first in over four decades in which an incumbent was defeated. Far from casting doubt on the nature of the politics of the Railroad Commission, these two elections provide persuasive evidence that organization and money, not incumbency, are decisive in determining who serves on the Commission.

The 1976 Campaign

Seven individuals entered the Democratic primary to fill Ben Ramsey's seat. Of these, three might have been expected to attract support from members of the petroleum industry. David Finney, a state representative from Fort Worth, Jon Newton, a state representative from Beeville, and Terence O'Rourke, a Houston attorney, were all more or less knowledgable about and favorably disposed toward the industry. It was therefore by no means a foregone conclusion that a single candidate would become the industry favorite and receive the bulk of its contributions.

The necessity of a united choice, however, was perceived early by the more politically aware members of the industry, for two other aspirants to the Democratic nomination were anathema to them. Jerry Sadler, a former Railroad Commissioner (1939–1942), former state representative, former state Land Commissioner, and an old-fashioned East Texas Populist, was widely regarded as a dangerous demagogue. Lane Denton, a new-fashioned consumer-oriented liberal from Waco, was openly hostile to the industry. The remaining candidates, Woodrow Wilson Bean from El Paso and Bob Wood from San Antonio, were not considered serious adversaries, but Sadler and Denton were acknowledged to be formidable antagonists. Industry activists, who saw that either man could win the primary if they were divided in their campaign efforts, set about building a consensus with determination.

Early in the winter of 1975, a number of state politicians and prominent private citizens began researching the background, opinions, and character of the proindustry candidates. This research consisted largely of making phone calls to mutual friends of theirs and the candidates, inquiring into their personalities, political views, and public records. Those active in sounding out their fellows included retiring Commissioner Ben Ramsey; the former United States ambassador to Australia, Ed Clark; and the executive vice-president of the Mid-Continent Oil and Gas Association, Bill Abington. As 1976 began, a consensus began to emerge in this group that Jon Newton was the most electable of the acceptable candidates. His record as chairman of the Texas House Energy Committee was reassuring, and he was well liked personally. By the end of January, this group began making phone calls to their politically active friends in the larger cities, urging them to think seriously about supporting Newton.

For his part, Newton adroitly took advantage of the informal network of power in Texas to advance his own cause. Late in 1975, he hired George Christian as a campaign adviser. A former aide to Lyndon Johnson, Christian had many friends among the politically influential in the state. Christian got Houston media expert Bob Heller involved in the Newton campaign.

Newton also benefited from his good relationship with Lieutenant Governor Bill Hobby. Hobby permitted one of his staff members, Tom Hagen, to manage Newton's campaign. Hagen was experienced in statewide campaigning and familiar with Hobby's list of friends and contributors throughout Texas.

From its inception, therefore, the Newton campaign was integrated into a network of informal contacts and personal relationships that guaranteed at least access to money and influence. Whether or not he would take advantage of this network and parlay it into electoral victory depended on his own judgment and skill, but simply being close to it so early was a great advantage possessed by no other candidate.

By February, the members of the early group had convinced their activist industry friends in the larger cities (especially Houston) that Newton was the most supportable candidate. These active industry members were all socially and politically prominent independent oilmen with extensive experience in this sort of campaigning. Many of them were associated with Mid-Continent or the Texas Independent Producers and Royalty Owners Association. These men drew up a list of eighty or so primary prospects for campaign contributions. All the people on the list were wealthy independents, friends and acquaintances who were used to relying on the smaller active group for early information about candidates. The activists began organizing informal get-togethers—largely cocktail parties—at which Newton was presented to small groups of these wealthy men, who could then independently form their own opinions of his character and politics.

It is important to realize that there was nothing foregone or automatic about these endorsements. Newton's introduction to the inner circle only gained him access to the fountain of campaign finances; it did not guarantee that he would be able to make it flow. If Newton had not been talented in interpersonal relations, or if he had made political blunders in his discussions with the people he met at these small receptions, their support would have gone elsewhere.

The fact is, however, that Newton was both personally impressive and politically astute. As he attended more and more of these industry sizing-up sessions, his campaign committees began to receive substantial contributions. An internal industry agreement was forming that Newton was the intelligent choice for the Railroad Commission, and a stream of money was the natural consequence.

Newton's contacts with potential supporters began in the metropolitan areas during the winter, but by spring he was meeting with groups in the smaller cities. As the wealthier and more active members of the industry began to adopt him as their favored candidate, the process of consensus building described in chapter 7 came into play. At the TIPRO convention in March,[10] at the American Institute of Mining Engineers meetings, at lunch in the various petroleum clubs, at the monthly Commission statewide hearings, in thousands of conversations across the state, the word was passed: Newton is a good man. He is electable. He needs money. As the industry made up its collective mind, the contributions rolled in.

Newton did not receive all the campaign contributions from the industry. Finney retained considerable backing in his hometown of Fort Worth, and O'Rourke also received scattered industry support. But Newton's exploitation of his personal opportunities was so successful that by the end of April he had received over a hundred fifty thousand dollars in contributions, compared to O'Rourke's fifty thousand and Finney's seventeen thousand.[11]

Newton did not rely solely on private persuasion and money to win for him. The Lo-Vaca issue (see chapter 5) was the focus of most public discussion during the campaign, and it was obvious that, if any candidate had been labeled anticonsumer, his ambitions for the nomination would have been hopeless. The press and the candidates assumed that the Railroad Commission's alleged failure to protect consumers was the key issue and acted accordingly.[12] In such a strategic environment, Newton's campaign was designed to present him as the consumer's friend but not as the enemy of the industry. His public statements were calculated to walk a tightrope between proindustry and proconsumer sentiments.

In many press releases and speeches during the spring of 1976, Newton sought to portray himself as a consumer champion. He attacked the Railroad Commission for laxity in monitoring Lo-Vaca's revenues.[13] He proposed that Coastal States, Lo-Vaca's parent company, be required to rebate money to customers.[14] He suggested that the state sales tax on utility bills be repealed.[15] He cited his proconsumer record in the Texas House, where he had voted for a refining tax and supported the creation of a state Public Utility Commission.[16] He set up a campaign steering committee in Bexar County (San Antonio), with a membership that included prominent liberals.[17] Although the other candidates, especially O'Rourke and Denton, attempted to portray him as anticonsumer,[18] Newton's public statements undercut the credibility of such charges.

In the absence of a specifically proindustry or anticonsumer candidate, name identification seems to have been the chief selector of the two top vote getters in the May primary. Jerry Sadler received a plurality of the vote, largely, it was assumed by all observers, on the strength of his well-known name.[19] Newton came in second, on the basis of the recognition he had been able to establish with media advertising. The success of Newton's strategy of presenting himself as a consumer candidate is suggested by the fact that he carried fifty of the fifty-four counties served by the Lo-Vaca system.[20]

The one-month runoff race between the two surviving candidates was much like the original primary campaign, with Sadler accusing Newton of being a tool of the industry—he told an Austin audience at one point that "if my opponent's suit were put in a wringer, they'd get enough oil to fill the crankcase of a Cadillac"[21]—and Newton relying on his enormous financial advantage to saturate the public consciousness with advertising. According to official figures, he outspent Sadler during the month of May by a ratio of eighty to one.[22]

The actual vote on June 5, however, was considerably different from the one in the primary. Newton's victory was substantial. He carried all but six rural counties near Sadler's East Texas home and received 66 percent of the popular vote.

But it did not have to be so. If the petroleum industry had been unable to concentrate its support and had instead split its contributions among Newton, Finney, and O'Rourke, all three might easily have done poorly, and Sadler might have won the primary outright.

As it was, Newton eventually received almost 78 percent of the total contributions to all seven candidates (see table 2). He was thus able to buy the exposure necessary to force Sadler into a runoff.

It should be obvious that this lack of industry fragmentation was not just a lucky break. Newton's victory was the product of careful organizing and political calculating, both within his campaign and within the industry. The pattern of industry-candidate interaction in the 1976 election is exactly the same as that observable in those elections where an incumbent defends a seat. This example strongly supports the contention that it is the consensus-building power of the petroleum industry, not the simple fact of incumbency, that has made industry candidates for the Railroad Commission invincible.

The 1980 Results

In the 1980 Democratic primary, a consumer-oriented, aggressively antioil candidate, Jim Hightower, lost narrowly to incumbent Jim Nugent. John Poerner became the first Railroad Commission incumbent since 1938 to lose his seat in an election, being defeated by State Representative Buddy Temple.

Even more than in the election of 1976, the results of the 1980 primary reinforce the conclusions drawn in the present chapter. In the race for both seats, the candidate that spent the most money emerged victorious. Temple was the first Commission challenger since the beginning of record keeping in Texas to outspend an incumbent, investing almost $600,000 in the race compared to Poerner's $474,000. And, in the traditional mode, Nugent successfully defended his seat by more than doubling Hightower's expenditures of about $210,000.[23]

Nor do the 1980 results detract from the present generalizations about industry electioneering. Temple courted petroleum support, making proindustry pronouncements to the press and endorsing energy interests on thousands of television spots.[24] In the end, about 16 percent of his contributions came from people associated with industries regulated by the Commission.[25] He won partly because his proindustry stance prevented petroleum activists from seeing him as a threat and vigorously opposing him. In contrast, the industry's reaction to the Hightower campaign verged on hysteria.[26] Although Hightower refused contributions from anyone associated with a Commission-regulated industry on principle, it is apparent that, even had he not done so, he would have received virtually nothing from anyone in the petroleum industry. This complete lack of support from oil and gas was an important factor behind the financial malnutrition that accompanied his campaign.[27]

THE FUTURE

If the first one of the decade is any guide, Democratic primaries in the 1980s should be considerably livelier than those of previous eras. Despite the more complicated setting, however, the basic facts seem to be unchanged: although cleavage between consumers and producers has arisen to play a significant role in its electoral campaigns, at bottom the politics of the Railroad Commission is still the politics of money.

The great fear of Jim Hogg that an elected agency would be subject to domination by the interest it was supposed to regulate seems to have been realized. This raises disturbing questions about democratic accountability, which will be addressed in chapter 9.

It might seem that, given the resources that will be available to members of the industry in the foreseeable future, industry candidates will surely continue to prevail in Railroad Commission elections for many years to come. But this conclusion is not necessarily consistent with the discussion in this chapter. Hightower's strong showing against Nugent underscores the fact that the politics of the Commission is evolving. It would be foolish to expect the future to be no different from the past. It is not impossible to imagine a situation in which a serious, ably led campaign by a popular liberal might attract sufficient financing to cancel out the industry candidate's advertising advantage. Or, equally likely, the Commission might become so important in the eyes of the public that its electoral campaigns become, in effect, high-stimulus races. Under those circumstances, the free publicity received by an antiindustry candidate might overbalance the advertising of the industry candidate. If or when either of these possibilities becomes a reality, a different sort of politician might emerge victorious from the Democratic primary.

Were that to happen, the logical countermove by industry activists would be to support the Republican nominee in the general election. Whether their informal organizational network would work as well within Texas' minority party as it has within the dominant party is impossible to say. But, whether or not they succeeded in defeating the Democratic nominee, they would undoubtedly rescue the Republicans from the hopelessness that has been the lot of their candidate in the past. A victory for an antiindustry candidate might thus be the motive force behind the emergence of the Republican party as a significant challenger in state politics.

Part 4

The Ironies of Politics

9. Lessons

I guess oil is about the only business left with any romance to it.
—A Railroad Commission field inspector, in conversation, April 1978

The room is a large one, perhaps the size of a high school gymnasium. Yet it is packed. Men, plus a sprinkling of women, all in expensive suits, sit shoulder to shoulder in rows in plastic chairs. There are several hundred of them. Around the periphery of the rows of chairs, other people, equally well tailored, stand and chat in groups. In the front of the room, there is a table with three empty chairs facing the assemblage. This table is bathed in light, for three television camera crews are setting up their equipment. The rows of chairs are divided by a central aisle, as in a church. At the head of this aisle, fifteen feet in front of the table, is a single microphone on a stand.

The setting is the banquet room of an expensive hotel in Austin. Gathered together are the top executive officers of every major integrated oil company in the country and many of the larger independents, plus their attendant lawyers, chief engineers, and public relations advisers. The era is the late 1970s. The occasion is the Railroad Commission annual state-of-the-industry hearing. The season is early spring. The mood is concern tinged with desperation.

When the Commissioners arrive, they sit in back of the table, facing the executives, and the chairman makes a short welcoming speech. Television cameras are focused, and the groups of conversationalists disappear. The chairman calls the first "witness" to the microphone, and the event begins.

One by one, the presidents or vice-presidents of the giants of the petroleum industry troop to the microphone to address the Commissioners, their fellow executives, and, they hope, the world outside. Each is supposed to speak for fifteen to twenty minutes, although it is not unusual for several to exceed this limit. The script they follow seems traditional. First, they summarize the activities and accomplishments of their company in the past year. Next, they assess the probable developments of the company and the industry in the coming year, paying special attention to the likelihood of an increase or a decrease in the supply of oil and gas. Finally, they denounce the federal government's meddling in the industry, praise the wise and restrained manner in which the Railroad Commission has regulated their business, and implore the nation to return to the free-market system as the only road to prosperity and energy sufficiency.

These people are successful and wealthy. They are accustomed to commanding thousands of employees on several continents. The public frequently imagines them as grand conspirators, ably manipulating the destinies of nations. Governments treat them practically as fellow sovereigns. There is every reason to expect them to be confident and capable as they stand at the microphone.

Yet the tone of their speeches is one not of sureness and mastery but of distress and bewilderment. The speakers are baffled over why the public is hostile to them. They are outraged by the federal government's insistence on knowing their secrets and overseeing their operations. They are anguished by the attacks of environmentalists and consumer groups. Most of all, they are dismayed that the American nation seems to be abandoning the free-market system.

All morning and half the afternoon, they walk to the front of the room and plead their case to the Commissioners and the TV cameras. "Bureaucracy is choking the petroleum industry," proclaims the vice-president of one of the largest firms in the world, "market pricing is the only system that's ever worked" in preventing oil shortages. The present system of price control is "government-mandated inefficiency," argues the president of another, and "we should return to the free-market system." The present regulations discourage exploration and enhanced recovery, says a third speaker. Repeal all environmental laws and return to market mechanisms, claims a fourth, and oil will again gush from the ground. Each states it differently, but each says the same thing: the country will go down the drain unless we abandon federal regulation and return to the free-market system, as in the good old days of the Railroad Commission.

The free-market system? The Railroad Commission? The Commission that supported American oil prices for forty years by keeping a lid on Texas production? The Commission that devised a system of spacing and allocation so contorted that it was the despair of the companies represented by many of these speakers? The Commission that forced producers to save natural gas when it was practically worthless? The Commission that directed pipelines to purchase oil and gas fairly rather than economically, that protected marginal wells from economic pressures, that constantly lobbied in Washington for import restrictions? This is the agency that is now being praised as the model for a free-market approach to national oil policy?

Yes, that Railroad Commission. All day, some of the most able and powerful executives in the world stand at the microphone and speak nonsense, while the TV cameras whir. Representatives of an industry that was sustained for over a generation by state interference in the market plead again and again for a return to the imagined freedom of a previous era. Then, they go home and no doubt wonder why the public regards them as hypocrites.

It would be tempting to explain this yearly outpouring of foolishness as a cynical public relations ploy by a group of people who believe that ordinary citizens are too stupid to catch on to their con game. But such an explanation will not do. To stand in that room and watch the desperate manner in which the executives give their speeches is to be convinced of their sincerity. They do not have the polished delivery of the political huckster; instead, they stammer and search for words to express their injured righteousness. It is obvious that they genuinely, even passionately, believe that it is the free market that has visited prosperity and power on their industry and their country. They authentically regard the Railroad Commission as a historical helpmate to an industry that actually needs little assistance. They really think that if regulations were lifted from them, and the golden age of free enterprise restored, their troubles would wither away.

In short, it is not contempt for the public but intellectual confusion that prompts the corporate executives to make such absurd statements about past and present oil policy. The ideology of the free market, plus a romanticized notion of its historical relation to their industry and the Railroad Commission, fogs their view of current events. Misperceiving the past, they misunderstand the present and grow ever more isolated and unable to make an intelligent case for their own survival.

The fact is that, as chapters 2, 3, and 4 of this book have demonstrated, from the 1930s to the 1970s the domestic petroleum industry was enmeshed in a web of state regulations specifically designed to shield it from the ravages of the market. This system was created at the behest of the majors; later, when its implications became clear, it was supported by the independents. The system was a creature of the state rather than of the federal government, but that does not make it a free-market system.

This system was run by persons actively sympathetic to the industry, rather than neutral or hostile, as was the case with the federal government in the 1970s. Apparently this is the source of the confusion in the minds of the speakers at the state-of-the-industry hearing. They apparently believe that the free market is good and that government regulation is bad. Under the tutelage of the Railroad Commission, they felt protected; the Railroad Commission must therefore be good. If it was good, it must by definition be part of the free-market system. The federal government is not sympathetic to them; it must therefore be bad. If it is bad, it must be contrary to the free-market system. The result of this half-conscious process of thought is the public display of Alice in Wonderland reasoning evident in petroleum industry speeches.

The danger of this misanalysis is that it misleads both the critics and the defenders of the Railroad Commission. If discussions of state regulation of the industry focus on the alleged value of the free-market system, then the Commission's real contributions are ignored. Because of the Commission, there was no free market for oil in Texas after 1935. But that is only the fundamental outline of the relationship of the state to the industry. There are more subtle and more important points to be made about the value of petroleum regulation. These deal with the importance to the industry and the nation of protecting the independent section of the industry from the assaults of unrestrained competition.

PETROLEUM AND CAPITALISM

The economic system that predominates in the United States is known as capitalism. Under this system, most economic decisions are made by private individuals pursuing their personal profit according to the dictates of market forces. Defenders of capitalism have always argued that submission to an unrestricted market forces individuals to act in a manner that ultimately benefits the public. Critics of capitalism have countered that the public interests are damaged by such uncoordinated economic decision making.

Although the executives at the state-of-the-industry hearing undoubtedly speak for their fellow members of the industry in endorsing the free-market view, in fact their business could barely survive under a completely free market. As was explained in chapter 2, the unrestrained pursuit of individual profit in oil production leads both to industry self-destruction and to enormous waste of the nation's resources. An unregulated market is consequently not a defensible mode of operation of the industry, despite propaganda to the contrary. Since the 1930s, the industry has willingly submitted to the coordinating and justice-dispensing authority of the Railroad Commission and other state regulatory agencies, all the while proclaiming its allegiance to the capitalist ideal.

The fact that the industry has been regulated for half a century, however, does not mean that it is not capitalist. Despite the state regulations, most of the decisions about investment, organization, growth, and competition in the industry are made privately. Although the Railroad Commission and related agencies have created an artificial environment in which the industry has functioned, within the special confines of that environment private choices have predominated.

In order to evaluate the place of the Railroad Commission in the history of the petroleum industry, it will be helpful to explore the effects of the Commission on the evolution of the industry as part of the capitalist system.

CAPITALISM AND INNOVATION

One of the important discussions of the capitalist system was authored by the Austrian economist Joseph Schumpeter in the 1940s. Capitalism, he argued, is based on constant innovation. Business entrepreneurs are artists in industry: they create new forms, new techniques, and new methods, thus causing the industry to progress. In Schumpeter's words,

> . . . the function of entrepreneurs is to reform or revolutionize the pattern of production by exploiting an invention or, more generally, an untried technological possibility for producing a new commodity or producing an old one in a new way, by opening up a new source of supply of materials or a new outlet for products, by reorganizing an industry and so on. . . . To undertake such new things is difficult and constitutes a distinct economic function. . . . To act with confidence beyond the range of familiar beacons and to overcome . . . resistance requires aptitudes that are present in only a small fraction of the population.[1]

In Schumpeter's view, innovation is essentially a nonrational act. Its source is the wellsprings of human creativity that are as yet barely understood by psychologists. Innovation in business is very similar to creativity in science, in art, or in human relations generally. It comes from a distinct personality type and is not duplicable by normal, routinized processes. This view of business innovation is supported by other observers. Elting Morison's review of the personalities of nineteenth-century inventors makes it clear that they more nearly resembled the stereotype of the bohemian artist than the stereotype of the business executive.[2] Arthur Koestler's examination of creativity in art, science, and humor offers convincing evidence that the capacity for innovation is a character trait possessed by some individuals, rather than an attribute of this or that mode of organization.[3] These two would most likely agree with Schumpeter that the progress of capitalism depends on its sustaining the entrepreneurial spark that lies at the root of its evolution.

Schumpeter does not believe that such sustenance is possible, however. Capitalism relies not only on entrepreneurs, he argues, but on organized rationality, on routine and order—in a word, on bureaucracy. He predicts that, as capitalist enterprises grow, they will become oppressively regimented, just as do government bureaus. Eventually, he argues, bureaucracy will choke off the creative impulse in capitalist enterprises, and the entrepreneurial spark will be extinguished. Without this nonrational source of growth, capitalist firms will become unimaginative and inflexible, innovation will cease, and industry will fail to advance. In the end, industries will be socialized by democratic governments responding to a public clamor to do something about this senile incapacity. Hence, in Schumpeter's argument, the advance of one of the main ingredients of capitalism—organization—will destroy the basis for its other main ingredient—innovation. The capitalist system will destroy itself with its own success.[4]

As an outstanding example of capitalist development, the petroleum industry would seem to be made to order for testing Schumpeter's predictions. Has the business of producing and selling oil and gas become dominated by large bureaucratic organizations that are incapable of adapting to changed circumstances? Is the petroleum industry doomed to strangle to death on rationality? Does the oil business contain the seeds of its own destruction?

The answer to these questions is no, at least partly because of the efforts of the Railroad Commission.[5] In large measure because the Commission has protected and nurtured its "irrational" and "inefficient" elements—that is, the independents and small landowners—by shielding them from market forces, the industry has retained its entrepreneurial spark and thereby sustained its vitality.

To suggest that independents were economically irrational and inefficient is not to accuse them of poor management practices or to imply that they were basically parasitic. The contention that most independents would not have survived a free market is not slander but recognition of their position within the structure of the industry. Because most independents were small, they lacked the economies of scale associated with the major integrated companies. Hence, they were forced to invest more capital per unit of petroleum produced. Additionally, because independents tended to be concentrated in the poorer parts of the poorer fields (also, especially, because they tended to own marginal wells), they were bound to invest more to produce less, even discounting their lack of economies of scale. Finally, because the great majority of independents produced domestically, they were unable to realize the advantages of importing cheap foreign oil. Irrespective of any *individual* talent for business, independents were therefore inefficient *as a group* because they had to invest more than the optimum for each barrel of oil they produced. From the standpoint of economics, they were irrational.

As argued in chapters 2 through 6, one of the major historical projects of the Commission has been to champion independent producers and small landowners in this hostile economic environment. The fundamental policy in this program was market-demand prorationing. By restraining production, supporting prices, and rationing participation in the market, the Commission ensured that the relatively inefficient independent companies would not be exterminated by competition from the major integrated companies. Additionally, in spacing and allocation decisions, protection of stripper wells, regulation of pipelines, and other similar policies, the Commission discriminated in favor of the independents and small landowners, thereby enabling them to survive.

The paradox of this historical mission of the Railroad Commission is seen in the fact that, by protecting the smaller companies of the industry from the rigors of the free market, the Commission performed a valuable service to the industry as a whole. Because the independents are the source of a significant proportion of the creative risk taking in the industry, their existence has been the foundation of much of its vigor. When it is viewed in a broad historical context, the Commission's protection of the Texas independents has saved the industry from decay and has thus contributed to the national interests.

The public views the industry as being dominated by enormous firms, often called the Seven Sisters.[6] But in fact, in both its oil and its gas wings, the petroleum industry is highly fragmented. Although huge firms dominate the marketing of petroleum products, they share production, transportation, and refining with thousands of smaller companies.[7] Compared to many other industries, petroleum is a happy hunting ground for small business.

The Railroad Commission can be given at least partial credit for this state of affairs. Always taking the side of the little firms in important policy questions, Commissioners created a set of economic circumstances favorable to small enterprise. While in such states as California and Montana, which did not prorate to market demand, independents were at the mercy of market forces, in Texas they were protected and nurtured. In Texas, as in no other state, the spirit of innovation survives in the petroleum industry, and the Railroad Commission is largely responsible.

Of course, protecting all those relatively inefficient producers in order to encourage those who might be creative involved a considerable amount of short-run waste. Most independents were not innovative, and those who weren't were supported along with those who were. But that is a small cost to pay for keeping the entrepreneurial spark alive. Besides, the cost to the nation of Commission regulations, as discussed in chapter 6, does not seem particularly onerous when surveyed from the perspective of the 1980s.

In order for this argument to be plausible, however, the value of the independents will have to be demonstrated. Just what do independents accomplish that makes them worth protecting? In brief, a strong case can be made to demonstrate that independents as a group are willing to take risks to a far greater degree than are integrated companies.

INDEPENDENTS AND RISK TAKING

Although Schumpeter tended to lump all sorts of innovation together, there are actually several categories of entrepreneurial activity. In general, at least four types of creative behavior can be observed in industrial activity:

1. Technological innovation: the introduction of a device or pro-

cess. This activity breaks down into several more, including invention, first application, and large-scale development.

2. Reorganization of a corporation's structure to adapt to changing economic demands. Some corporations blunder along from decade to decade, relying on an outmoded structure to attempt to deal with economic circumstances; some reorganize themselves and become more efficient.[8]

3. Organization of the flow of money. Creativity can be applied to financial organizations as well. Business empires have been created that rested largely on the efforts of one individual to manipulate finances.[9]

4. Taking risks. In some business ventures, the probability of success is small, but the potential rewards are very large. Some individuals are willing to take the chance of meeting with disaster, in order to experience also the possibility of making a fortune.

When the possible types of innovation are broken down in this way, it becomes clear that modern corporations, as exemplified by the Seven Sisters of the petroleum industry, are well adapted to some and poorly adapted to others. Almost by definition, for example, corporate reorganization and the widespread application of a new technology are the province of big companies.

The majors have also received credit in the scholarly literature on innovation for their inventions and first applications of these inventions, although there may be some cause for doubt here.[10] Research on patterns of technological innovation usually lumps innovation with commercial development, so a company that first uses a new technique on a broad scale receives credit for first thinking of it. This is entirely plausible, since the majors maintain huge research and development departments. Some independents claim privately, however, that many technical innovations are invented by independents and subsequently widely applied by a major company, which then gets the credit.[11] In the absence of hard evidence to the contrary, majors must be considered responsible for most technological innovations in the industry, but a research project specifically aimed at disentangling research from development might increase the credit given to the independents.

The real province of the independents, however, lies in the fourth area of innovation: risk taking. As enormous corporations, the majors are responsible to stockholders, who demand a steady, predictable return on their investment. Furthermore, the bureaucratic structure of the majors fosters a reliance on routine predictability. As a result, one of the characteristic attributes of a major company is caution. As John Kenneth Galbraith has remarked of the contemporary corporation,

> . . . the development of the modern business enterprise can be understood only as a comprehensive effort to reduce risk. It is not going too far to say that it can be understood in no other terms.[12]

It is not that the majors take no chances. Every business venture, especially every oil venture, involves some uncertainty. Large corporations are not suited, however, to displaying truly significant amounts of initiative or daring. Majors almost always follow current conventional wisdom; not for them are unorthodox theories or eccentric intuitions. Normally, when they apply a new technique or discover a new field, it is because there is abundant evidence that the undertaking is relatively safe. The gambles, the hunches, the personal visions, the offbeat theories, and the great risks they leave to the independents.

For it is the function of independents in the petroleum industry to rush in where angels fear to tread. Most independents are relatively sober executives; some are famous kooks. But all chose their line of work because they prefer not to be bound by traditional thinking and bureaucratic timidity. For stability and responsibility in the petroleum industry, the nation relies on the majors, but for daring and imagination it must look to the independents.

At the root of the petroleum industry is the discovery of oil. Yet there is no known method of finding it directly, before the drill is put into the soil. Oil explorers use a variety of techniques for inferring the location of the resource, from the employment of sophisticated electronic equipment to reliance on pure guesswork, but, whatever the method, every new well is still a considerable gamble. About 40 percent of all wells drilled in the United States are dry, and about 90 percent of the wells drilled by wildcatters are dry.[13] However much major companies try to systematize and make predictable their operations, they cannot avoid the fact that the discovery of their basic resource is a highly problematic task.

The majors do find oil; statistics suggest that they have discovered slightly more than half the nation's reserves.[14] The oil credited to the majors, however, is usually the safe oil, found where contemporary scientific theory suggests a strong possibility that the drill will meet with success.

In other words, as a general rule majors discover the low-risk oil. But, because scientific knowledge of the occurrence of petroleum has always been highly imperfect, real advances in exploration have historically been the province of companies willing to take large risks. And here the independents shine.

The history of petroleum exploration is studded with examples of important discoveries made by independents, working against accepted theories in areas condemned by major companies as being without productive potential. Spindletop and East Texas are just two famous examples of independents outpacing the majors; such incidents continue.[15]

These historical achievements dramatically illustrate the value, in the business of finding petroleum, of sustaining a group of audacious mavericks. Their contribution to the industry as a whole is illustrated by the conclusions of a survey conducted by the *Oil and Gas Journal* in 1963. The journal staff investigated the discoveries of the giant oil fields in the United States, that is, those that contained one hundred million barrels or more. Of the 241 giant fields found up to then, 122, or 50.6 percent, had been discovered by independents; 111, or 46 percent, had been discovered by majors; and 8 were uncreditable. Fifty percent of the country's yearly production, 50 percent of its reserves, and 60 percent of its probable ultimate recovery were expected to come from these fields.[16]

Within Texas, the contribution of the independents is even more striking. In 1976, the twelve largest oil fields in the state accounted for almost 54 percent of its recoverable reserves. Of these fields, independents discovered six, accounting for 39.11 percent of the state's reserves, and majors discovered five, accounting for 11.72 percent. One was discovered jointly.[17] A very substantial proportion of the nation's petroleum, then, has been added by the section of the industry that is most willing to take exploration risks.

There are also, however, less famous and more pedestrian kinds of risk taking at which independents also excel. Once oil and gas are discovered, their transportation out of the ground does not follow as a matter of course. Each reservoir is unique: the permeability and porosity, the presence or absence of faulting, the kind of drive present, and many other technical aspects of the field can affect its recovery rate. As scientific knowledge of recovery has advanced, new techniques for bringing the petroleum to the surface have been discovered and developed. Because every field is unique, however, the application of a new method of recovery or the application of an old method in a new field always involves some magnitude of risk. The cost of failure can be high. There has been no systematic research on this point, but anecdotal evidence suggests that, once again, independent companies often play a leading role in testing new recovery techniques and thus advancing the state of the art of the industry as a whole.

For example, the South Ward field in Ward County north of Fort Stockton was discovered in 1929. By the late 1940s, its production had fallen off to such an extent that all its wells had become marginal. In 1949, a medium-large independent, the Forest Oil Company, took over production in the South Ward field from Standard of Texas (a subsidiary of Standard of California—one of the largest majors).

One of the secondary recovery techniques which was beginning to be widely used in the 1940s is known as waterflooding. In a waterflood project, water is forced down one or several wells (patterns vary) in order to increase pressure in the reservoir, thereby boosting production in the remaining wells.

In the late 1930s and the 1940s, waterflooding became popular in North and Central Texas. Despite its potential for increasing production, however, the industry hesitated to apply it in West Texas. There were several reasons for this reluctance, the most important being that hydrogen sulfide gas was associated with the oil in most West Texas reservoirs. This "sour gas" was corrosive, and operators had trouble keeping their casings from disintegrating in the "sour fields" of the area. They were afraid that the oxygen in the water involved in a waterflood would worsen the problem.

But Forest wanted to experiment. It farmed out a lease in South Ward from Standard and applied to the Railroad Commission for a permit to waterflood. "We were willing to take those risks," as one of Forest's executives explained in 1980, "in order to expose ourselves to success, so to speak." After hearings, the Commission issued a permit for the project.[18]

In the next five years, Forest extracted more oil from the South Ward field than had been produced there in its first twenty years of primary recovery.[19] More important, with this pathbreaking example in front of them, the major companies were no longer reluctant to try waterflooding projects in some of their holdings in West Texas. Forest had demonstrated that the combination of hydrogen sulfide and fresh water was not as damaging as the other companies had feared. Gulf, Mobil, and several other companies began secondary recovery in the Permian Basin, to the benefit of the industry and the nation.

Forest was not finished, however. Until the early 1950s, all waterflooding had been attempted in sandstone formations. Although oil occurs in other kinds of permeable rock, their relative unfamiliarity had deterred experimentation. But in 1954 Forest decided that it was worth the risk to attempt a waterflood project in its holdings in a dolomite limestone formation in the South Cowden field of Ector County, near Odessa. After determining that success in this enterprise would require rather close well spacing, Forest applied to the Railroad Commission for a permit to waterflood and, in addition, an exemption from the state spacing rule.

This time there was active opposition from one of the majors. Stanolind (now Amoco), convinced that such a project was foolish, objected to the spacing in Commission hearings. Commissioner Murray sat in on these hearings and became convinced that Forest's project deserved support. The permits were issued.[20] Once again, Forest's gamble paid off: South Cowden's productivity was greatly increased. Stanolind was so impressed that it eventually joined Forest in the project, unitizing their holdings for mutual benefit.

In 1974, Forest sold its leases in South Ward and South Cowden to the Sun Oil Company. They were still producing and still profitable. But, again, the greater value of the project lay in its demonstration to the rest of the industry that secondary recovery was feasible in dolomite limestone formations. By helping itself, Forest had helped the nation.

This is only one example of an independent leading the industry in technological advance, but conversations with petroleum producers suggest that it is not atypical. All progress involves a willingness to "take those risks in order to expose ourselves to success." If the United States petroleum industry has progressed rapidly—and it has—this is at least partly because it contains a group of individuals and companies more concerned with opportunities than with dangers.

It consequently appears that a new twist must be added to the economic theories of Joseph Schumpeter. The onward march of capitalist ossification may have proceeded apace in other industries, but in the American petroleum industry it has been held at bay by policies explicitly intended to perpetuate the smaller, "irrational" part of the business. The entrepreneurial spark has been kept glowing, as it were, by the Railroad Commission. In the short run, this has meant that the domestic industry has not approximated efficient development, but in the long run it may well have secured the industry's survival. The Commission must be awarded much of the credit for preserving the romance of the oil business and thus ensuring its vitality.

THE ADVANTAGES OF SMALL SIZE

Willingness to take risks is not the only useful characteristic of independents, however. The very fact that most of them are quite small relative to the majors enables them to fulfill certain functions inappropriate to huge corporations. As the insects in a forest fit into an ecological niche different from but no less vital than that of the large mammals who inhabit the same area, the smaller independents fit into an ecological niche less grandiose but no less important than that of the integrated companies.

Majors have very great resources and a large corporate structure. They therefore tend to explore areas that promise a large return on their investment. They are capable of drilling in areas (such as offshore) where costs are so high as to be beyond the reach of any but the largest independents, but they often neglect the lesser possibilities closer to home. Some oil fields in the United States cover just forty acres; the great majority contain only a million barrels of oil or less.[21] Many pockets of gas are similarly tiny. Majors seldom drill where the evidence indicates such a paltry accumulation. They consequently tend to discover only larger fields, which make their investment worthwhile.[22]

A collection of petroleum beneath the majors' notice, however, is often enough to sustain many independents. Some independents subsist on smaller, relatively poor fields or on tiny pockets of remaining gas or oil in a field already developed and abandoned as depleted by a major company. While each of these small companies individually adds very little to the nation's production, as a group they make a very considerable contribution.

Logue and Patterson is an example of one of these independents. This company specializes in finding small gas pools in already developed areas of Texas. Company employees examine geologic and electric logging to identify underground structures where minor pockets of gas might be located. In 1978 and 1979, Logue and Patterson drilled nine producers (out of fifteen wells—an unusually high success ratio) in four different gas fields in South and East Texas. None of these wells is prolific, but Logue and Patterson does not require great quantities of gas to operate profitably.

Like many smaller companies, Logue and Patterson also picks up the scraps left by the majors. In 1975, its executives discovered that Sunray (now Sunoco) had abandoned several of its wells in the Heyser field on the Gulf coast. Logue and Patterson bought the lease and cleaned out the casings in the idle wells. It turned out that Sun had not noticed that there were shallow petroleum-bearing rock zones above the depth at which it was producing. Logue and Patterson began to recover oil from one of these strata and gas from another. In the five years after the old wells were rejuvenated, the company produced 900,442 thousand cubic feet from the gas well. The oil well contributes eight barrels a day.[23] Again, neither of these wells is going to solve the energy crisis, but multiplied across Texas they add significantly to the nation's reserves.

MARGINAL WELLS AND THE NATIONAL INTERESTS

Of all the Railroad Commission efforts to protect the little people of the petroleum industry, its custodial care of stripper wells raised a historical furor second only to its discrimination in favor of small tracts. In the days of market-demand prorationing, the Commission, under statutory obligation, gave first priority to marginal wells before it determined what total level of allowable production remained to allocate to the state's prolific fields. For decades, the wells capable of producing many relatively cheap barrels of oil each day were suppressed so that wells producing two, or eight, or fifteen barrels a day of expensive oil would have a guaranteed market.

This situation moved economists to great indignation. The vision of tens of thousands of barrels of inexpensive oil going unproduced because of solicitude for those hordes of tiny wells laboriously pumping expensive oil seemed self-evidently absurd to them. "The oil that is *not* produced is invariably the oil that could be produced at the lowest marginal cost," wrote Homan and Lovejoy.[24] "We cherish and fertilize these weeds at the expense of the flowers," concurred Morris Adelman.[25] For most of the 1950s and 1960s, these attacks continued.

In the short run, this objection to strippers was justified. The support of marginal wells undoubtedly raised United States oil prices and perhaps served to discourage exploration for flush fields (see chapter 6). From the perspective of the 1980s, however, these costs take on a different light. When a well is drilled, the mud used to lubricate the bit is forced out into the oil-bearing strata, where it clogs the tiny pores in the rock. In the industry, this is known as formation damage. In a newly opened field, the pressure in the reservoir is so great that it unclogs the formation, pushing the mud out of the rock and back up the well shaft. A marginal well is marginal, however, precisely because there is very little pressure remaining underground. If an abandoned stripper is redrilled, formation damage will usually block the passage of whatever oil remains, and the well will not flow.

Moreover, many strippers are operated in fields that contain underground water. By the time a well reaches the marginal stage, it may produce much more water than oil. In old fields like those around Luling, for example, it is not uncommon for marginal wells to make one barrel of oil and a hundred barrels of water a day. (The Commission, of course, requires operators to dispose of this water in an appropriate manner—usually by reinjection into the formation.) When such a well is plugged, the underground water floods the formation, closing off the channels of oil. If the well is later redrilled, and the problem of formation damage can be overcome, it will nevertheless produce only water.

In other words, the great majority of marginal wells must be kept pumping, or their potential production will be lost forever. When economists advocated abandoning strippers in the 1950s and 1960s, they were recommending the permanent loss of the reserves under those wells.

Thousands of marginal wells are pumping today because fifteen and thirty years ago the Commission ignored economists and nurtured these weeds at the expense of the flowers. The economic argument assumed that abundant sources of cheap oil would always be available from foreign countries. Taking that for granted, economists could see no justification for supporting strippers on the off chance that the nation might one day need their production.

With the dislocation of the 1970s, however, the United States was thrown back on its own resources. That we had as rich a resource base as we did was partly to the credit of the marginal wells and the Railroad Commission. In 1976, strippers accounted for 11 percent of the production and about 20 percent of the reserves in Texas.[26] In the long run, it paid the nation to preserve marginal production all those years, and the long run has now arrived.

THE RAILROAD COMMISSION AND THE SMALL PRODUCER: A SUMMARY STATEMENT

Certainly because they thought it was in the interests of the Texas economy, and perhaps because they thought it was in the national interests, Railroad Commissioners for decades directed a system of regulation that buffered the small producer and landowner from market forces. By supporting prices, rationing the market, and favoring the industry's little people, they created an economic haven within the state for the insects of the industry. This haven had its costs, as described in chapter 6. There is no denying that, from the standpoint of maximum efficiency or, as economists say, "the optimum allocation of resources among alternative employments,"[27] it was irrational.

But there are other things worth pursuing besides economic rationality. By helping support the independent segment of the industry, the Commission aided in preserving its entrepreneurial spark. By promoting the industry's little people, it ensured that every possible source of petroleum would be exploited to the fullest. By fostering marginal wells, it preserved a great resource into the 1980s. These actions were the despair of economists at the time, but history has made them look quite justifiable. In retrospect, the Commission's irrationality seems more like wisdom.

This conclusion is one that cannot be reached by the sort of executives who gather at the state-of-the-industry hearings, however, because of their almost pathological inability to come to terms with the meaning of government regulation. Fearing regulation in the abstract, and hating the federal government, they are beyond understanding that intelligent regulation can be their best friend.

In the context of the political battles over national energy policy in the 1970s and 1980s, this ideological blindness cripples the industry's ability to plausibly defend its own positions on issues. It also deprives the public of important information, for, if the true historical relationship between the industry and the government is unknown, the country lacks guideposts for choosing sensible national policy. Debate over government regulation takes place in a twilight zone between outright propaganda and cherished myth, and we are prevented from learning from the past. If the petroleum industry is crippled by government intervention because of public exasperation over industry behavior, its members will have brought their defeat upon themselves.

THE PROBLEM OF DEMOCRATIC CONTROL

When examined in a long-run perspective, the substance of Railroad Commission policy making gets fairly high marks. There is another aspect of politics on the Commission, however: method as opposed to substance, process as opposed to outcome. In a democratic country such as the United States, a supremely important industry like petroleum might be expected to be under the influence of national democratic control. Yet an examination of the politics of the Railroad Commission is not comforting to partisans of democratic accountability.

In the first place, the men who have exercised such influence over the supply and price of United States energy for half a century have been almost completely independent of the influence of citizens who have lived outside Texas. Until the 1970s, an occasional appearance before a committee of Congress was the only notice Commissioners were forced to take of a national constituency. Commission policies significantly affected the lives of the inhabitants of many other states, yet these citizens had only the frailest influence over any Commission decision.

By passing the Connally Hot Oil Act in 1935, the United States Congress delegated oil regulatory power to the states, of which Texas was incomparably the most important. From that year to 1973, Commissioners were free to cultivate a state rather than a national constituency. That the choices they made under these conditions were fairly responsible does not alter the fact that they were severed from the authority of most of those whose lives they would directly affect.

In the 1970s, domination of the national oil situation shifted to a location still farther removed from democratic control: OPEC. Because the countries of this organization have been far more aggressive in their determination to wring advantages from the United States consumer than the states that practiced market-demand prorationing ever were, the essential similarity of the two situations escapes notice. In fact, however, the actions of OPEC in controlling production to *raise* prices are entirely comparable to the actions of Texas and the other market-demand states to *support* prices. In both cases, administrative agencies acting to control supply, and thus price, have made decisions in an atmosphere divorced from a consideration of the opinions of consumers. OPEC does not pretend to be democratically accountable, but it is not composed of United States citizens. The Railroad Commission has no such defense.

In the second place, even within Texas the structure of state politics has combined with the structure of the petroleum industry to deflect democratic influences. As discussed in chapters 7 and 8, the political clout of the independent producers is quite sufficient to dominate state policy making in this one vital area. The absence of credible party competition has made it difficult for countervailing tendencies to Commission policies to find expression in electoral challenges. The informal organizational web of the independent producers has enabled them to reach a consensus on candidates and coordinate efforts on their behalf. The abundant wealth at their disposal has enabled them to supply their chosen candidates with overwhelming campaign resources.

The charge does not have to be made that Commissioners have been bought or captured by the independents; the problem is one not of personal integrity but of public access. The independent producers have not had to buy Commissioners because they have been able to ensure that only individuals favorable to their interests would attain Commission authority. In general, it is not too great an exaggeration to say that what has existed in Texas for nearly fifty years is not public regulation of the petroleum industry but regulation of the industry as a whole, under public auspices, by the independent portion of that industry.

This sort of situation is by no means uncommon in American politics. Many researchers have concluded that government regulatory agencies tend to be dominated by the industries they regulate.[28] In the case of these other agencies, however, their personnel had been appointed. It might have been expected that an elective agency would be different. The only difference between the Railroad Commission and other regulatory agencies, however, appears to be that one segment of the industry, and the economically weaker segment at that, dominates in Texas, whereas in other agencies the industry as a whole or its stronger components seem to be dominant. If the Railroad Commission can be said historically to have been under democratic control, it is because the industry's numerous small companies have been its major constituency rather than the few large firms.

It would appear that Governor Hogg's fears have indeed been borne out. The penchant that Commissioners have for resigning in midterm, however, illustrates a fact that makes Hogg's distinction between an elective and an appointive agency largely irrelevant. The governor is also favorably disposed toward the independents and makes appointments accordingly. Under these conditions, it is idle to speculate about whether an appointive or an elective Commission would come closer to approximating the democratic ideal.

The Administrative Procedures Act and the Open Meetings Act, passed by the state legislature in the mid 1970s in an effort to deal with the problem of the private application of power in a public agency, forbid Commissioners to discuss Commission cases in any but a public forum. While well intentioned, however, such laws cannot prevent the private, nonofficial fraternizing of Commissioners and Commission staff with members of the industry. Again, the problem is not that Commissioners allow themselves to be bribed or browbeaten by their acquaintances in the industry. Rather, constant association is bound to affect their perspective and their sympathy. Any Commissioner knows hundreds of people who personify industry interests—but who represents public interests? Commissioners would answer that *they* do, but such a claim is not convincing.

If the problem of personal contact between the industry and Commissioners is basically intractable, the problem created by campaign contributions is even more so. Huge advantages in financing have made industry candidates invulnerable to challenges in the past. In the future, they may not be invulnerable, but they will remain formidable. It must make the partisans of democracy uneasy to behold a political environment in which the naked power of money has been so plainly decisive in the electoral arena.

It is hard to imagine, however, any reform that would improve this situation. Federal courts made it clear in the 1970s that they would not tolerate a law that forbade private campaign contributions.[29] Even if they did, however, the difficulty would only shift, not disappear. Since the crucial election is the Democratic primary, there would be no party resources available to candidates. If private donations were prevented, then the overwhelming advantage in primaries would go to those candidates who could supply their own resources—that is, the rich. It is impossible to believe that such a change would improve either the caliber of the Commission's policy making or the quality of its democratic accountability.

The only remaining possibility would be to have the federal government take over the job of regulating Texas' oil and gas fields. This scheme has great advantages in terms of democratic theory, for it would mean that the Texas industry, which is of vital national importance, would at last be under national control.

Whatever appeal this suggestion has in theory, however, it has none when examined in the light of experience. Far from being closer to popular influences than the Railroad Commission, federal bureaucracies have demonstrated conclusively that they habitually put their own organizational interests over public interests.[30] Furthermore, whereas the Railroad Commission has displayed tolerable efficiency in administration, the federal government's track record with, among other things, natural gas pricing regulation and the strategic petroleum reserve does not inspire confidence in its ability to control energy production. In brief, a federal take-over of the Texas petroleum industry would probably involve all the disadvantages of regulation by the Railroad Commission yet offer none of the advantages.

It would seem, therefore, that the state and the nation should anticipate a future Railroad Commission that operates much as it has in the past.

CONCLUSION

This has been a book about choices. For half a century, the choices made in a small and obscure state agency have directed many of the activities of some of the world's corporate giants and exerted an often decisive influence over United States energy policy. The study of the forces underlying the Railroad Commission's regulation of oil and gas in Texas has provided an opportunity to understand not only the nature of the political process in the United States but also the origins of our current national energy quandary.

Chapter 3 of this book emphasized the ironies that accompanied the regulation of the petroleum industry by the Railroad Commission. In this final chapter, however, a greater irony has become evident. For half a century, Texas Railroad Commissioners have chosen petroleum policies that have been, in the main, intelligent and public-spirited. Yet they have made these choices in a political atmosphere dominated by private rather than public concerns and subject to a constituency that barely includes the broad citizenry that must live with their decisions. The substance of Commission policies has been generally admirable, but the process that produced them must be disappointing for friends of democracy.

As the era of energy scarcity lengthens, the Railroad Commission will probably become a less important actor in the national reaction to the crisis. But the problems posed by the paradox of the Commission's politics will not diminish. Whether the setting is Texas or the United States, there is still much to be said about petroleum, politics, and public policy.

Notes

1. The Railroad Commission of Texas

1. *Information Please Almanac* (New York: Information Please Publishing Company, 1978), pp. 48–49.
2. *Statistical Abstract of the United States, 1934,* p. 688; 1938, p. 739; 1943, p. 755; 1951, p. 696; 1952, p. 705; 1966, p. 715; 1978, p. 754.
3. Fred Pass (ed.), *Texas Almanac, 1978–1979* (Dallas: A. H. Belo, 1977), p. 406.
4. Marver H. Bernstein, *Regulating Business by Independent Commission* (Princeton: Princeton University Press, 1955), pp. 86–92, 157–160, 170, 184–185, 254–255; Louis M. Kohlmeier, Jr., *The Regulators* (New York: Harper and Row, 1969), pp. 69–82.
5. M. M. Crane, "Recollections of the Establishment of the Texas Railroad Commission," *Southwestern Historical Quarterly* 50 (1947): 478–486; Robert C. Cotner, *James Stephen Hogg: A Biography* (Austin: University of Texas Press, 1959), pp. 217–218, 223–241.
6. Graham T. Allison, *Essence of Decision* (Boston: Little, Brown, 1971).
7. Richard Rose, *What Is Governing?* (Englewood Cliffs, N.J.: Prentice-Hall, 1978).
8. Richard C. Snyder, H. W. Bruck, and Burton Sapin (eds.), *Foreign Policy Decision-Making* (New York: Free Press of Glencoe, 1962); Robert A. Dahl, *Who Governs?* (New Haven: Yale University Press, 1961), pp. 332–333.
9. Harold Lasswell, *Politics: Who Gets What, When, How* (New York: World, 1958).
10. James W. McKie, "Market Structure and Uncertainty in Oil and Gas Exploration," *Quarterly Journal of Economics* 74, no. 4 (November 1960): 547.
11. *Oil Directory of Texas, 1978* (Austin: R. W. Byram, 1978), p. 194.

2. Foundations, 1930–1935

1. John Stricklin Spratt, *The Road to Spindletop: Economic Change in Texas* (Austin: University of Texas Press, 1970), pp. 61–110.

2. John D. Hicks, *The Populist Revolt* (Lincoln: University of Nebraska Press, 1959), pp. 147–167, 433–439.
3. Cotner, *James Stephen Hogg*, pp. 147–167.
4. Alwyn Barr, *Reconstruction to Reform: Texas Politics, 1876–1906* (Austin: University of Texas Press, 1971), p. 120.
5. Spratt, *Road to Spindletop*, p. 216.
6. James A. Clark and Michel T. Halbouty, *Spindletop* (New York: Random House, 1952), p. 60.
7. Carl Coke Rister, *Oil! Titan of the Southwest* (Norman: University of Oklahoma Press, 1949), p. 412.
8. Arthur M. Johnson, *Petroleum Pipelines and Public Policy, 1906–1959* (Cambridge, Mass.: Harvard University Press, 1967), pp. 113–115.
9. Erich W. Zimmermann, *Conservation in the Production of Petroleum* (New Haven: Yale University Press, 1957), p. 145.
10. Ibid.
11. James A. Clark and Michel T. Halbouty, *The Last Boom* (New York: Random House, 1972), p. 109.
12. James Presley, *A Saga of Wealth* (New York: G. P. Putnam's Sons, 1978), p. 136.
13. Clark and Halbouty, *Last Boom*, pp. 114–116.
14. Henrietta M. Larson and Kenneth Wiggins Porter, *History of Humble Oil and Refining Company* (New York: Arno, 1976), pp. 397–402.
15. Presley, *Saga*, p. 128.
16. Economists do not agree among themselves whether the oil industry is self-adjusting. For an argument that parallels my own, see Paul H. Frankel, *Essentials of Petroleum* (London: Chapman and Hall, 1946), p. 67. For an argument that the industry *is* self-adjusting, see Morris A. Adelman, *The World Petroleum Market* (Baltimore: Johns Hopkins University Press, 1972), pp. 13–44. For a discussion that is more equivocal in its conclusions, see Edith Penrose, *The International Petroleum Industry* (Cambridge, Mass.: M.I.T. University Press, 1968), pp. 165–171.
17. Stephen L. McDonald, *Petroleum Conservation in the United States* (Baltimore: Johns Hopkins University Press, 1971), pp. 24–25, 76–84.
18. Ibid., pp. 85–86; Morris A. Adelman, "Efficiency of Resource Use in Crude Petroleum," *Southern Economic Journal* 31, no. 2 (October 1964).
19. Robert E. Hardwicke, "The Rule of Capture and Its Implications as Applied to Oil and Gas," *Texas Law Review* 13 (1935).
20. J. A. O'Connor, Jr., "The Role of Market Demand in the Domestic Oil Industry," *Arkansas Law Review* 12, no. 4 (Fall 1958): 345.
21. Larson and Porter, *Humble*, pp. 454–455, 472–473; Bennett H. Wall and George S. Gibb, *Teagle of Jersey Standard* (New Orleans: Tulane University Press, 1974), p. 254.
22. James A. Clark, *Three Stars for the Colonel* (New York: Random House, 1954), p. 9.

23. *East Texas Oil*, June 1934, p. 71.
24. Clark and Halbouty, *Last Boom*, p. 124.
25. Rister, *Oil!* pp. 317–318.
26. Clark, *Three Stars*, p. 11; Presley, *Saga*, p. 147.
27. Clark, *Three Stars*, p. 11.
28. Zimmermann, *Conservation*, pp. 121–135.
29. Rister, *Oil!* pp. 297–299; Clark and Halbouty, *Last Boom*, p. 149.
30. John C. Calhoun, Jr., *Fundamentals of Reservoir Engineering* (Norman: University of Oklahoma Press, 1976), pp. 201–203.
31. Clark and Halbouty, *Last Boom*, pp. 190–193.
32. Ibid.
33. Ibid., pp. 191–195.
34. *East Texas Oil*, May 1934, pp. 48–49.
35. For example, see H. C. Weiss (president of Humble), "Some Current Problems in Oil Conservation," address before the annual dinner, Petroleum Section of the American Institute of Mining and Metallurgical Engineers, New York City, February 16, 1939, pp. 7–8. Copy in possession of the author.
36. Acts, 36th Leg., reg. sess., 1919, chap. 155.
37. Acts, 41st Leg., reg. sess., 1929, chap. 313.
38. Acts, 42d Leg., 1st C.S., 1931, chap. 26.
39. Acts, 42d Leg., 4th C.S., 1932, chap. 2.
40. This chronology relies heavily on Robert E. Hardwicke, "Legal History of Conservation of Oil in Texas," in *Legal History of Conservation of Oil and Gas: A Symposium* (Chicago: American Bar Association, 1938).
41. *Macmillan et al.* v. *Railroad Commission*, 51 F. 2d 400 (W.D. Tex. 1931).
42. See note 39.
43. Rister, *Oil!* p. 321.
44. *Constantin et al.* v. *Lon Smith et al.*, 57 F. 2d 227 (1932).
45. *People's Petroleum Producers, Inc.* v. *Smith*, 1 F. Supp. 361 (1932).
46. See note 39.
47. *Amazon Petroleum Corp.* v. *Railroad Commission*, 5 F. Supp. 633 (S.D. Tex. 1934); *Amazon Petroleum Corp.* v. *Railroad Commission*, 5 F. Supp. 639 (E.D. Tex. 1934), considered on other grounds, 293 U.S. 388 (1935).
48. Eldon Stephen Branda (ed.), *The Handbook of Texas: A Supplement*, vol. 3 (Austin: Texas State Historical Association, 1976), p. 642; *Austin American*, October 22, 1947.
49. Texas State Archives, Railroad Commission Collection, boxes 2-10/568, 4-3/329.
50. George Sessions Perry, *Texas: A World in Itself* (Gretna, La.: Pelican, 1975), p. 171.
51. Clark and Halbouty, *Last Boom*, p. 217.

52. *Fort Worth Star-Telegram*, September 30, 1935.
53. Clark and Halbouty, *Last Boom*, p. 217.
54. *Time*, July 5, 1934, p. 50; *East Texas Oil*, May and June 1934.
55. Larson and Porter, *Humble*, p. 465.
56. Ibid., p. 464.
57. Two typical examples are J. Edward Jones, "A Memorial to the Members of the Congress of the United States Regarding Problems of the Petroleum Industry" (New York: Hamilton, 1934), Texas State Archives, Railroad Commission Collection, box 4-3/331, and F. W. Fischer, "The Independent's Viewpoint," *East Texas Oil*, August 1934, p. 74.
58. *New Outlook*, June 1934, pp. 22, 24.
59. Clark and Halbouty, *Last Boom*, pp. 155, 168.
60. Presley, *Saga*, p. 140.
61. Larson and Porter, *Humble*, p. 464; Clark, *Three Stars*, p. 79.
62. *Handbook of Texas*, p. 1004.
63. Clark, *Three Stars*, pp. 41–43, 47–64.
64. *Kilgore Daily News*, September 23, 1935.
65. Harold L. Ickes, *The Secret Diary of Harold L. Ickes: The First Thousand Days, 1933–1936* (New York: Simon and Schuster, 1953), p. 30; Kendall Beaton, *Enterprise in Oil: A History of Shell in the United States* (New York: Appleton-Century-Crofts, 1957), pp. 385–386.
66. There is a copy of this letter in the Texas State Archives, Railroad Commission Collection, box 4-3/329. For Ickes' rather ambiguous account, see his *Diary*, p. 15.
67. Interviews; Presley, *Saga*, pp. 146–149; Johnson, *Pipelines*, pp. 222–223.
68. Ernest O. Thompson, "History of the Connally Act," Texas State Archives, Railroad Commission Collection, box 4-3/319.
69. Clark and Halbouty, *Last Boom*, pp. 214–223.
70. Presley, *Saga*, pp. 153–157.
71. In April of 1934 Charles Roeser, a Fort Worth–based independent, and Russell Brown, general counsel for the Independent Petroleum Association of America, testified at a congressional hearing in favor of the Thomas bill; in November of 1934, J. Edgar Pew of Sunray (now Sunoco, a major company) testified against it. See the Texas State Archives, Railroad Commission Collection, box 4-3/331, and the *Dallas Morning News*, November 20, 1934.
72. Texas State Archives, Railroad Commission Collection, box 4-3/331.
73. Texas State Archives, Railroad Commission Collection, box 4-3/329.
74. This speech is reprinted in Samuel B. Pettengill, *Hot Oil* (New York: Economic Forum, 1936), pp. 251–257.
75. Larson and Porter, *Humble*, p. 485.
76. *East Texas Oil*, December 1934, pp. 5–6; Clark, *Three Stars*, pp. 113–114.

77. Clark and Halbouty, *Last Boom*, pp. 157, 159.
78. *East Texas Oil*, May 1934, p. 2.
79. Norman E. Nordhauser, *The Quest for Stability: Domestic Oil Regulation, 1917–1935* (New York: Garland, 1979), p. 162.

3. The Ironies of Regulation, 1935–1950

1. McDonald, *Petroleum Conservation*, p. 152.
2. Wallace F. Lovejoy and Paul T. Homan, *Economic Aspects of Oil Conservation Regulation* (Baltimore: Johns Hopkins University Press, 1967), pp. 168–169; McDonald, *Petroleum Conservation*, pp. 184–185.
3. *Oil and Gas Journal*, March 2, 1950, p. 28.
4. *TIPRO Reporter*, January 1950, p. 6; *Oil and Gas Journal*, January 26, 1950, p. 163.
5. Adelman, "Efficiency," p. 105. Also see Myron W. Watkins, *Oil: Stabilization or Conservation?* (New York: Harper and Brothers, 1937), p. 34; James W. McKie and Stephen L. McDonald, "Petroleum Conservation in Theory and Practice," *Quarterly Journal of Economics* 76, no. 1 (February 1962): 99, 105; Lovejoy and Homan, *Oil Conservation*, pp. 8–32, 55–56.
6. Lovejoy and Homan, *Oil Conservation*, pp. 75–78; McDonald, *Petroleum Conservation*, pp. 201–209.
7. Larson and Porter, *Humble*, pp. 489–493; Beaton, *Shell*, pp. 379, 381–383.
8. Lovejoy and Homan, *Oil Conservation*, p. 80.
9. Ibid., pp. 73–75.
10. Elton M. Hyder, Jr., "Some Difficulties in the Application of the Exceptions to the Spacing Rule in Texas," in *Oil and Gas Law* (reprints from *Texas Law Review*) (Austin: Texas Law Review, 1954), pp. 497–501.
11. Railroad Commission internal document on the evolution of Rule 37. Copy in possession of the author.
12. Kenneth Culp Davis and York Y. Willbern, "Administrative Control of Oil Production in Texas," *Oil and Gas Law*, pp. 1008–1011, 1035–1036.
13. Olin Culberson, "Biography" file, Barker Texas History Center, Austin.
14. Gerald Forbes, *Flush Production* (Norman: University of Oklahoma Press, 1942), pp. 92, 189.
15. Larson and Porter, *Humble*, p. 493.
16. Hardwicke, "Legal History," p. 256.
17. York Y. Willbern, "Administrative Control of Petroleum Production in Texas," in Emmette S. Redford (ed.), *Public Administration and Policy Formation* (Austin: University of Texas Press, 1956), p. 43.
18. Robert E. Hardwicke, "Texas, 1938–1948," in Blakeley M. Murphey (ed.), *Conservation of Oil and Gas: A Legal History, 1948* (Chicago: American Bar Association, 1949), pp. 480, 490–491.

19. *Smith* v. *Stewart*, 68 S.W. 2d 627 (Tex. Civ. App. 1934); *Railroad Commission* v. *Magnolia Petroleum Co.*, 130 Tex. 484, 109 S.W. 1d 967 (1937).
20. Hardwicke, "Texas, 1938–1948," p. 491.
21. *People's Petroleum Producers, Inc.* v. *Smith*, 1 F. Supp. 361 (1932).
22. Interviews; Weiss, "Problems in Oil Conservation," p. 13; Hardwicke, "Texas, 1938–1948," pp. 479–481.
23. Robert E. Hardwicke, "Oil-Well Spacing Regulations and the Protection of Property Rights in Texas," *Oil and Gas Law*, p. 1890.
24. Adelman, "Efficiency," p. 104.
25. Acts, 42d Leg., reg. sess., 1931, chap. 50, art. 6049, pp. 534–535.
26. Figures on stripper wells and production come from the National Stripper Well Survey for relevant years, supplied by the Interstate Oil Compact Commission. Texas production figures are supplied by the Mid-Continent Oil and Gas Association. Number of wells comes from *Petroleum Facts and Figures, 1950* (New York: American Petroleum Institute, 1951), p. 131.
27. Hardwicke, "Texas, 1938–1948," p. 489.
28. *Railroad Commission* v. *Humble Oil and Refining Co.* (Hawkins case), 193 S.W. 2d 824 (1946), p. 832.
29. Eugene V. Rostow, *A National Policy for the Oil Industry* (New Haven: Yale University Press, 1948), p. 53.
30. *Texas Almanac, 1978–1979*, p. 410.
31. Zimmermann, *Conservation*, p. 238.
32. Ibid., pp. 237–246.
33. Ernest O. Thompson, "Flare Gas Wastage in Texas: Steps Taken to Utilize," speech to the American Gas Association, May 1, 1947, Texas State Archives, Railroad Commission Collection, box 4-3/318.
34. John R. Stockton, Richard C. Henshaw, Jr., and Richard W. Graves, *Economics of Natural Gas in Texas* (Austin: University of Texas, Bureau of Business Research, 1952), pp. 17, 27–35, 72–75; Beaton, *Shell*, p. 502.
35. Maurice Cheek, "Legal History of Conservation of Gas in Texas," in *Legal History*, p. 279.
36. Stockton, Henshaw, and Graves, *Gas in Texas*, pp. 231–236; Zimmermann, *Conservation*, p. 244.
37. Railroad Commission internal memo analyzing the Murray Committee report, in possession of the author, p. 1; Cheek, "Conservation of Gas," p. 279.
38. *Energy Information Digest*, Subcommittee Print, 95th Cong., 1st sess., Committee Print 95-71, stock no. 052-070-04305-9 (Washington, D.C.: Government Printing Office, 1977), p. 33.
39. 1899 Tex. Gen. Laws chap. 49, no. 3; Tex. Rev. Civ. Stat. 6008, 6014 (1925); Barth P. Walker, "What Is an Oil Well? What Is a Gas Well? What Difference Does It Make?" *Proceedings of the Fourteenth Annual Insti-*

tute on Oil and Gas Law and Taxation (Albany: Southwestern Legal Foundation, 1963), pp. 175–232.

40. Tex. Rev. Civ. Stat. Ann. art. 6008, no. d, c (Vernon).
41. *Clymore Production Co.* v. *Thompson*, 11 F. Supp. 791 (W.D. Tex. 1935); *Clymore Production Co.* v. *Thompson*, 13 F. Supp. 469 (W.D. Tex. 1936).
42. *History of Petroleum Engineering* (Dallas: American Petroleum Institute, 1961), p. 862; Texas Railroad Commission, *Annual Report of the Oil and Gas Division, 1942* (Austin: 1942), p. 49.
43. Cheek, "Conservation of Gas," p. 279.
44. Rister, *Oil!* p. 281; Acts, 43d Leg., reg. sess., 1933, chap. 100, p. 222.
45. Acts, 44th Leg., reg. sess., 1935, chap. 120; Cheek, "Conservation of Gas," pp. 281–284.
46. *Consolidated Gas Utilities Co.* v. *Thompson*, 12 F. Supp. 462 (W.D. Tex. Oct. 1935).
47. George H. Fancher, Robert L. Whiting, and James H. Cretsinger, *The Oil Resources of Texas: A Reconnaissance Survey of Primary and Secondary Reserves of Oil* (Austin: Texas Petroleum Research Committee, 1954), pp. 70, 112, 210, 289, 310.
48. Jack K. Baumel, "The Feasibility and Possibility of a Statewide Plan of Gas Proration and Ratable Take," paper presented at a meeting of the Interstate Oil Compact Commission, December 9–11, 1946, p. 8. This document is in the possession of the author.
49. Ibid., p. 9.
50. Railroad Commission, *Annual Report of the Oil and Gas Division, 1943* (Austin: 1943), p. 41.
51. *Fort Worth Star-Telegram*, December 22, 1944.
52. *Oil and Gas Journal*, November 10, 1945, p. 56.
53. *Oil Weekly*, December 24, 1945, p. 32; *Corpus Christi Caller*, December 19, 1945; *Austin American*, December 19, 1945.
54. *Oil Weekly*, March 5, 1945, p. 30; *Oil and Gas Journal*, February 9, 1946, pp. 56–58; March 2, 1946, p. 48.
55. Hardwicke, "Texas, 1938–1948," pp. 457–458.
56. *Oil and Gas Journal*, August 18, 1945, p. 109; November 10, 1945, p. 56; January 5, 1946, p. 45; February 2, 1946, p. 99; April 26, 1947, p. 85; May 3, 1947, p. 36.
57. *Texas Almanac, 1978–1979*, p. 410.
58. *Oil and Gas Journal*, June 16, 1945, p. 76; February 16, 1946, p. 76.
59. *Austin Statesman*, December 8, 1946.
60. Railroad Commission Oil and Gas Docket no. 129, order no. 4-10,351, March 17, 1947; *History of Petroleum Engineering*, p. 914.
61. *Oil and Gas Journal*, June 28, 1947, p. 91.
62. *Railroad Commission* v. *Shell Oil Co.*, 206 S.W. 2d 235 (1947).
63. *Railroad Commission* v. *Flour Bluff Oil Co.*, 219 S.W. 2d 506 (Tex. Civ. App. 1949), error ref'd, p. 508.

64. The major controversy after 1949 occurred over the Spraberry field. See *Railroad Commission* v. *Rowan Oil Co.*, 259 S.W. 2d 173 (1953), and Nelson Jones, "The Spraberry Decision," *Oil and Gas Law*, p. 2093.
65. I am not alone in holding this opinion. See Stockton, Henshaw, and Graves, *Gas in Texas*, p. 233.
66. Ibid., p. 233; Hardwicke, "Texas, 1938–1948," p. 485.
67. *Oil and Gas Journal*, January 26, 1946, p. 301.

4. The Balance Wheel, 1950–1965

1. Robert E. Sullivan (ed.), *Conservation of Oil and Gas: A Legal History, 1958* (Chicago: American Bar Association, 1960), p. 218.
2. *Dallas News*, February 12, 1953; *Texas Almanac, 1978–1979*, p. 410.
3. McKie and McDonald, "Theory and Practice," p. 110; *Dallas News*, June 27, 1948.
4. Texas State Archives, Railroad Commission Collection, boxes 4-3/319, 4-3/329, 4-3/320; Ernest O. Thompson, "How the Railroad Commission of Texas Works" (Austin: n.d.). Copy in possession of the author. *Houston Post*, October 11, 1959.
5. Clark, *Three Stars*, p. 231.
6. Ibid.
7. Davis and Willbern, "Administrative Control of Oil Production," pp. 1008–1011. Although this study was originally published in the 1940s, all contemporary observers agree that its principles also apply to the Commission of the 1950s.
8. Alfred Cameron Mitchell, *Market Demand and the Proration of Texas Crude Petroleum*, Texas Industry Series 10 (Austin: University of Texas, Bureau of Business Research, 1964), pp. 13–29.
9. Zimmermann, *Conservation*, pp. 142–144, 167–170.
10. Ibid., p. 142; McKie and McDonald, "Theory and Practice," p. 110.
11. Sullivan, *Legal History*, p. 237.
12. Ibid., p. 236.
13. McKie and McDonald, "Theory and Practice," p. 114.
14. See the memorandum from the chief examiner, Oil and Gas Division, to the Commissioners, June 16, 1959, p. 2 (Port Acres memorandum).
15. Charles Issawi and Mohammed Yeganeh, *The Economics of Middle Eastern Oil* (New York: Praeger, 1962), p. 91.
16. Carl Solberg, *Oil Power* (New York: New American Library, 1976), p. 200.
17. Ibid., pp. 192–193.
18. Harold F. Williamson, Ralph L. Andreano, Arnold A. Daum, and Gilbert C. Klose, *The American Petroleum Industry, 1899–1959: The Age of Energy* (Evanston, Ill.: Northwestern University Press, 1963), p. 815.
19. *Oil and Gas Journal*, March 21, 1960, p. 101.
20. Ibid., March 19, 1962, p. 102; March 26, 1962, p. 88.
21. Robert Engler, *The Politics of Oil* (Chicago: University of Chicago Press,

1961), pp. 230–247, 359–360; Williamson et al., *American Petroleum Industry*, p. 815.

22. *Dallas News*, October 29, 1958; Engler, *Politics of Oil*, p. 231; Jim Langdon, statement before the House Ways and Means Committee, oil import hearing, June 27, 1968; Langdon, "The U.S. Energy Crisis, Why?" address to the 44th annual meeting of the New Mexico Oil and Gas Association, Albuquerque, October 2, 1972, p. 4.
23. *Oil and Gas Journal*, January 18, 1960, p. 41; January 25, 1960, p. 127.
24. Adelman, "Efficiency," p. 104; McKie and McDonald, "Theory and Practice," p. 115.
25. *Oil and Gas Journal*, August 31, 1959, p. 36.
26. *World Oil*, August 1, 1959, pp. 102–108; *Oil and Gas Journal*, January 8, 1960, p. 41; July 2, 1962, p. 57.
27. Johnny Mitchell, "A License to Steal" and "Is This Well Necessary?" addresses given to multiple audiences prior to 1961, private files of Johnny Mitchell; Michel T. Halbouty, *Ahead of His Time* (Houston: Gulf, 1971), pp. 253–269.
28. *Oil and Gas Journal*, April 30, 1962, p. 43.
29. Halbouty, *Ahead*, p. 260.
30. Letter to three Commissioners from Johnny Mitchell, July 24, 1959, and Commissioners' reply, July 27, 1959, private files of Johnny Mitchell.
31. Railroad Commission internal memo, September 20, 1957. Copy in possession of the author.
32. Railroad Commission internal memo, June 16, 1959. Copy in possession of the author.
33. Port Acres decision: *Halbouty* v. *Railroad Commission of Texas et al.*, 357 S.W. 2d 364 (1962), p. 371.
34. Normanna decision: *Atlantic Refining Co. et al.* v. *Railroad Commission*, 346 S.W. 2d 801 (1961).
35. Port Acres decision.
36. Normanna decision, pp. 807, 810; Port Acres decision, pp. 369, 374–375.
37. Port Acres decision, p. 371.
38. Ibid., p. 376.
39. Ibid., p. 371.
40. *Petroleum Week*, March 31, 1961, p. 3; *Oil and Gas Journal*, April 30, 1962.
41. Jim Langdon, "The Influence of Court Decisions upon Railroad Commission Policy in Rule 37 Cases and the Allocation of Allowables to the Small Tract Well," address given to industry audiences on numerous occasions, 1963 and 1964. Copy in possession of the author.
42. *Railroad Commission* v. *Shell Oil Co.*, 380 S.W. 2d 566 (1964).
43. Acts, 59th Leg., reg. sess., 1965, chap. 303, art. 6008c.
44. McDonald, *Petroleum Conservation*, p. 176; Lovejoy and Homan, *Oil Conservation*, p. 152.

45. Solberg, *Oil Power*, p. 168.
46. J. E. Brantly, *Rotary Drilling Handbook*, 4th ed. (New York: Palmer, 1948), pp. 264–286.
47. *Alphonzo E. Bell Corp.* v. *Bell View Oil Syndicate* (24 Cal. App. 2d 587, 76 P. 2d 167) (1938).
48. Presley, *Saga*, p. 128.
49. Halbouty, *Ahead*, p. 271; *Oil and Gas Journal*, July 2, 1962, p. 57.
50. *Dallas News*, June 22, 1962; April 6, 1963.
51. Ibid., August 28, 1962.
52. William J. Murray, Jr., "Memorandum Regarding Directional Drilling in the East Texas Field," Railroad Commission internal memo, July 2, 1962, p. 5; hereinafter, this document will be identified as the Murray Memo. Copy in possession of the author.
53. *Dallas News*, July 13, 1963.
54. *Dallas Times-Herald*, August 25, 1967.
55. *Austin American-Statesman*, August 9, 1964.
56. *Oil and Gas Journal*, January 25, 1960, p. 126; May 7, 1962, p. 80.
57. Ibid., May 21, 1962, p. 106.
58. *Petroleum Facts and Figures, 1961* (New York: American Petroleum Institute, 1962), p. 226.
59. *Austin American-Statesman*, August 9, 1964; *Dallas Times-Herald*, August 25, 1967.
60. *Dallas News*, May 23, 1964; Clark and Halbouty, *Last Boom*, p. 275.
61. Acts, 56th Leg., reg. sess., 1959, chap. 23, pp. 598–599.
62. William J. Murray, Jr., "Scandal or Survival," address to the Mid-Continent Oil and Gas Association, Dallas, October 9, 1962, p. 4. Copy in possession of the author.
63. Much of the information in the following discussion comes from "A Review of Commission Actions Relating to Straight Holes, with Particular Attention to Those Involving Rule 37 Exceptions in the East Texas Field," Railroad Commission internal memo, July 17, 1962. Copy in possession of the author.
64. The earliest slant-hole clauses that I have been able to discover occur in April and May of 1947. For a representative, single-paragraph order, see Rule 37, permit no. 36,267 to Bobby Manziel on May 14, 1947. For a longer, more detailed example of a straight-hole clause, see Rule 37, permit no. 36,369 to L. W. Powell on August 8, 1947.
65. Frank Maloney, "Legal Aspects in Connection with Directional Drilling Cases," address, September 21, 1962, preserved in the Texas State House Library, 11.
66. See the various letters, orders, and court records in Railroad Commission East Texas folder 6-19638.
67. *Dallas Herald*, May 27, 1963; *Houston Post*, August 26, 1962.
68. Clark and Halbouty, *Last Boom*, p. 273.
69. Murray Memo, p. 8.

70. Ibid., pp. 14, 25; *Dallas News*, August 28, 1962.
71. Murray Memo, p. 10a.
72. *Dallas News*, April 12, 1962.
73. Murray Memo, p. 12.
74. Clark and Halbouty, *Last Boom*.
75. *Dallas News*, April 21, 1962.
76. *Oil and Gas Journal*, June 11, 1962, p. 116; July 9, 1962, p. 86; *Dallas News*, July 15, 1962; August 18, 1962.
77. *Dallas News*, August 1, 1962; February 23, 1963.
78. Ibid., June 7, 1962; June 10, 1962; June 15, 1962; June 23, 1962; June 19, 1963.
79. Ibid., August 28, 1962; July 11, 1963.
80. Robert MacAvoy, *The Crisis of the Regulatory Commissions* (New York: W. W. Norton, 1970), pp. 1–52.
81. *Dallas News*, May 23, 1964; Clark and Halbouty, *Last Boom*, pp. 275, 281.
82. *Dallas News*, May 23, 1964; *Dallas Times-Herald*, August 25, 1967.
83. *Dallas News*, May 27, 1962; June 17, 1962; June 21, 1962; June 29, 1963; February 11, 1964; *Austin American-Statesman*, August 9, 1964.
84. *Dallas News*, April 30, 1964; *Dallas Times-Herald*, August 25, 1967.
85. *Dallas News*, April 8, 1963; April 10, 1963.
86. *Austin American-Statesman*, June 19, 1963.
87. Supplement to *House Journal*, 53d Leg., reg. sess., April 14, 1953, pp. 145–151.

5. Welcome to the Energy Crisis, 1965–1980

1. Langdon, "Rule 37 Cases and Allocation."
2. *Dallas News*, December 14, 1977.
3. Morris A. Adelman, *The Supply and Price of Natural Gas* (Oxford: Basil Blackwell, 1962), p. 72.
4. *Texas Almanac, 1978–1979*, p. 410.
5. Adelman, *Natural Gas*, p. 37; *Oil and Gas Journal*, January 14, 1963, p. 48; *Corpus Christi Caller*, January 31, 1977.
6. *Thompson* v. *Consolidated Gas Utilities Co.*, 300 U.S. 55, 57 S. St. 55 (1937).
7. *Corzelius et al.* v. *Harrell*, 143 Tex. 509, 186 S.W. 2d 961 (1945).
8. W. Earl Ainsworth, "Ratable Take of Gas: A Pipeliner's Viewpoint," address to the Engineering Committee of the Interstate Oil Compact Commission, December 6, 1956, Miami Beach, p. 3. Files of TIPRO.
9. John S. Cameron, Jr., "How Texas Prorates Gas," address to the Engineering Committee of the Interstate Oil Compact Commission, December 6, 1956, Miami Beach. Files of TIPRO.
10. *Phillips Petroleum Co.* v. *State of Wisconsin*, 347 U.S. 672, 74 S. Ct. 794, 98 L. Ed. 1035 (1954).
11. *Northern Natural Gas Co.* v. *Corporation Commission of Kansas*, 372

U.S. 84, 83 S. Ct. 646, 9 L. Ed. 2s 601 (1963); *Oil and Gas Journal*, February 25, 1963, pp. 61–63.

12. *TIPRO Reporter*, February 1960, pp. 14–15; October 1962, p. 29; December 1962, pp. 15–18.
13. Dee Kelly, "Gas Proration and Ratable Taking in Texas," *Texas Bar Journal*, December 1956, pp. 796–797; W. L. Bowser, *Proration News and Trends*, March 6, 1959, pp. 1–2. Files of TIPRO.
14. *Oil and Gas Journal*, April 8, 1963, p. 50; *TIPRO Reporter*, February and March 1963, p. 5.
15. *Texas Almanac, 1978–1979*, p. 410.
16. Ibid., p. 404.
17. William J. Murray, Jr., testimony before the Subcommittee on Energy Regulation of the Committee on Energy and Natural Resources, Subtitle A, Title IX, S. 1308, Energy Supply Act of 1979, p. 11.
18. *Dallas Times-Herald*, August 30, 1978.
19. Railroad Commission Oil and Gas Docket no. 20, order no. 66,679, February 17, 1977.
20. Transcript of "Face the Nation," February 5, 1978, p. 6.
21. *Austin American-Statesman*, March 5, 1978.
22. *Daily Texan*, June 13, 1978.
23. Railroad Commission press release, February 8, 1978, p. 4.
24. Gas Proration Seminar program, Railroad Commission, pp. 10–20, 83.
25. Private tape recording of question-and-answer session between Railroad Commissioners and members of the industry, TIPRO annual convention, Houston, June 1978.
26. Letter from William J. Murray to Railroad Commissioners, regarding Federal Energy Regulation Commission order no. 46, September 20, 1979.
27. *Energy Information Digest*, p. 155.
28. *Texas Almanac, 1978–1979*, p. 410.
29. *Oil and Gas Journal*, January 28, 1963, pp. 235–239.
30. Paul Burka, "Power Politics," in *The Best of Texas Monthly, 1973–1978* (Austin: Texas Monthly Press, 1978), p. 316.
31. Ibid., p. 322. San Antonio's "City Public Service Board Reply to Coastal States' Letter and 'Report' Dated May 11, 1973," p. 3. This document comes from the files of the Austin city attorney's office; henceforward, it will be designated CPS.
32. *Dallas News*, April 16, 1976.
33. *Forbes*, January 1, 1974, p. 189.
34. Ibid.
35. Burka, "Power Politics," p. 323.
36. "O'Rourke Report on the Texas Railroad Commission's Failure to Regulate Lo-Vaca Natural Gas 'Banking' Deals," 1976; O'Rourke campaign document. Copy in possession of the author.
37. *New York Times Magazine*, June 5, 1966.

38. *Houston Post*, November 1, 1960; May 29, 1963; *Texas Parade*, January 1965, p. 36.
39. *Vernon's Texas Civil Statutes*, vol. 2A, art. 1119 (St. Paul: West, 1962), p. 665.
40. *Railroad Commission of Texas et al.* v. *Houston Natural Gas Corp.* (Alvin case), 155 Tex. 502, 289 S.W. 2d 559 (1956).
41. Number of employees from personnel records of the Railroad Commission. Number of gas utilities from Railroad Commission Gas Utilities Division *Annual Report* for 1972, p. 77.
42. CPS, pp. 29–30; *Austin American-Statesman*, March 19, 1978.
43. CPS, p. 29.
44. *Wall Street Journal*, June 22, 1973.
45. *Austin American-Statesman*, March 19, 1978.
46. Ibid. *Prospectus*: "Percentage Interests in a Securities Trust and a Gas Search Trust Pursuant to an Offer of Settlement by Coastal States Gas Corporation and Coastal States Gas Producing Company," p. 4. This document has several sections, without collation of pages. Hereafter, it will be designated as *Prospectus*. The reader should beware of page number duplications.
47. *Dallas News*, March 29, 1973.
48. *Prospectus*, p. 5.
49. *San Antonio Express*, September 29, 1973; June 28, 1974; *San Antonio Light*, February 19, 1975.
50. *San Antonio Express*, June 16, 1974; June 23, 1974; *Austin American-Statesman*, June 19, 1974.
51. *Port Arthur News*, February 8, 1977; AFL-CIO news release, March 23, 1976; *Austin American-Statesman*, August 14, 1976.
52. *San Antonio Express*, January 11, 1974; July 18, 1974.
53. *Houston Chronicle*, April 1, 1974.
54. *Austin Citizen*, March 22, 1974.
55. *Houston Chronicle*, February 27, 1977.
56. *Austin American-Statesman*, April 2, 1977; *Houston Post*, April 2, 1977; *San Antonio Express*, May 3, 1977.
57. *Vernon's Annotated Revised Civil Statutes of the State of Texas*, vol. 3, art. 1446c (St. Paul: West, 1979), pp. 308–310. See also *Southwestern Bell Telephone Co.* v. *Public Utility Commission of Texas et al.*, 571 S.W. 2d 503 (1978).
58. *Prospectus*, p. 13.
59. Ibid., cover page.
60. *Dallas Times-Herald*, December 18, 1977; *Beaumont Enterprise-Journal*, December 26, 1977; *Austin American-Statesman*, December 13, 1977.
61. *San Antonio Express*, December 13, 1977; *Dallas Times-Herald*, December 13, 1977; *Austin American-Statesman*, December 13, 1977.
62. *Austin American-Statesman*, December 13, 1977.

63. *San Antonio Express*, August 29, 1978; *San Antonio Light*, September 9, 1979.
64. *San Antonio Light*, September 9, 1979; *Prospectus*, pp. 14–17.

6. Policy and Its Consequences

1. Zimmermann, *Conservation*, pp. 279–290.
2. *Rules and Regulations of the Texas Railroad Commission, Oil and Gas* (Austin: R. W. Byram, 1958); see also *Texas Oil and Gas Handbook* (Austin: R. W. Byram, 1978).
3. *Pollution vs. the People*, Joint Report of the Interim Committee on Pipeline Study and Beaches, 62d Legislature of the State of Texas, 1971, pp. 1, 8–10.
4. "Texas: The Superstate," *Newsweek*, December 12, 1977, pp. 34–46; Marlan Blissett, Bob Davis, and Harriet Hahn, "Energy Policy in Texas: State Problems and Responses," *Public Affairs Comment* 21, no. 3 (May 1975): 2.
5. Campaign speech of Ernest O. Thompson, August 8, 1936. Copy in possession of the author.
6. Jim Langdon, "The Future of Market Demand Proration in Texas," address to the American Petroleum Institute, Houston, January 3, 1967, p. 2. Copy in possession of the author.
7. This estimate comes from the Texas Independent Producers and Royalty Owners Association. No federal or state agency has apparently attempted another estimate.
8. Adelman, "Efficiency," p. 107.
9. Mack Wallace in *Texas Oil Journal* 45, no. 7 (November 1978): 8. Statement of Jim Langdon to the Commerce Committee, United States Senate, April 22, 1974, p. 12. Copy in possession of the author. Several statements by Ernest Thompson, 1930s, Texas State Archives, Railroad Commission Collection, box 4-3/329.
10. California data were obtained from the Conservation Committee of California Oil Producers, Los Angeles, in table form. Texas data are from the *Oil Directory of Texas* for 1960 to 1978.
11. Testimony of L. Frank Pitts before the Senate Finance Committee, June 25, 1979.
12. *Daily Texan*, October 26, 1979; *Austin American-Statesman*, October 28, 1979.
13. Zimmermann, *Conservation*, pp. 291–293.
14. Adelman, "Efficiency"; McKie and McDonald, "Theory and Practice"; Lovejoy and Homan, *Oil Conservation*, pp. 265–276.
15. Adelman, "Efficiency," pp. 105, 116–122.
16. Lovejoy and Homan, *Oil Conservation*, pp. 190–193; John M. Blair, *The Control of Oil* (New York: Pantheon, 1976), p. 181.
17. *Petroleum Facts and Figures, 1963* (New York: American Petroleum Institute, 1964), pp. 88, 152, 231.

18. Ibid., pp. 234, 235.
19. *Oil and Gas Journal,* January 13, 1969, p. 44.
20. Adelman, "Efficiency," pp. 116, 122.
21. *Petroleum Facts and Figures, 1971* (Washington, D.C.: American Petroleum Institute, 1972), p. 469.
22. *Daily Texan,* February 13, 1980.
23. Adelman, "Efficiency," p. 106.
24. Parts of this argument are available in the speeches of Commissioner Jim Langdon. See his address to the 22d annual meeting of TIPRO, San Antonio, May 28, 1968, pp. 9–11, and to the 44th annual meeting of the New Mexico Oil and Gas Association, Albuquerque, October 2, 1972, pp. 3–4.
25. Franklin M. Fisher, *Supply and Costs in the U.S. Petroleum Industry* (Baltimore: Johns Hopkins University Press, 1964), pp. 3, 35.
26. Ibid., p. 4.
27. Peter Bachrach and Morton S. Baratz, "Two Faces of Power," *American Political Science Review* 56 (1962).
28. McDonald, *Petroleum Conservation,* pp. 197–226.
29. Rostow, *National Policy,* p. 119; Halbouty, *Ahead,* pp. 232–238.
30. There is no survey evidence to confirm that Texans as a group are unusually attached to the concept of private property. Historical and anthropological research, however, supports this contention. See T. R. Fehrenbach, *Lone Star* (New York: Macmillan, 1968), pp. 707–709, and Evon Z. Vogt, "American Subcultural Continua as Exemplified by the Mormons and Texans," *American Anthropologist* 57 (1955).
31. *Weekly Oil News,* May 27, 1976, p. 1.

7. Influence

1. Bernstein, *Regulating Business,* pp. 157–160; William L. Cary, *Politics and the Regulatory Agencies* (New York: McGraw-Hill, 1967), p. 67.
2. David B. Truman, *The Governmental Process* (New York: Alfred A. Knopf, 1951), pp. 264, 321–332; Norman J. Ornstein and Shirley Elder, *Interest Groups, Lobbying and Policymaking* (Washington, D.C.: Congressional Quarterly, 1978), p. 83.
3. See the list of production totals of the six thousand or so producers in the state in the *Oil Directory of Texas, 1978,* pp. 1–222. See also Halbouty, *Ahead,* pp. 59–69.
4. *Texas Revised Civil Statutes* (Vernon's Supplement, 1979), art. 6252, 13A and 17, pp. 242, 271.
5. Bernstein, *Regulating Business,* pp. 185–186.
6. *Texas Oil Report,* June 22, 1961, p. 1.
7. Truman, *Governmental Process,* pp. 306, 355, 419–420; Ornstein and Elder, *Interest Groups,* pp. 54–83.
8. *Texas Oil Journal* 45, no. 2 (June 1978): 3.
9. Truman, *Governmental Process,* pp. 139, 148, 189, 190, 210; Abraham

Holtzman, *Interest Groups and Lobbying* (New York: Macmillan, 1966), p. 14.

10. Joseph A. Schlesinger, "The Politics of the Executive," in Herbert Jacob and Kenneth N. Vines (eds.), *Politics in the American States* (Boston: Little, Brown, 1965), p. 65.
11. *Houston Post*, April 16, 1963; January 8, 1965.
12. *Daily Texan*, January 9, 1951; *Houston Post*, November 1, 1960; *Dallas News*, August 18, 1961.

8. Campaigns and Elections

1. Richard H. Kraemer and Charldean Newell, *Texas Politics* (St. Paul: West, 1979), pp. 97–98.
2. Blissett, Davis, and Hahn, "Energy Policy in Texas," p. 2.
3. Demographic data are from the U.S. census of 1970. Production data are from the Mid-Continent Oil and Gas Association, Dallas.
4. Warren E. Miller and Teresa Levitin, *Leadership and Change* (Cambridge, Mass.: Winthrop, 1978), pp. 30–40.
5. Dan Nimmo and Robert L. Savage, *Candidates and Their Images* (Pacific Palisades, Calif.: Goodyear, 1976), pp. 13–39; Donald E. Stokes and Warren E. Miller, "Party Government and the Saliency of Congress," in Angus Campbell, Philip E. Converse, Warren E. Miller, and Donald E. Stokes, *Elections and the Political Order* (New York: John Wiley, 1967), pp. 204–209.
6. These financial records are based on raw data supplied by the Enforcement Division of the Texas Secretary of State's office.
7. *Oil Directory of Texas, 1978–1979* (Austin: R. W. Byram, 1977).
8. John J. McCloy, *The Great Oil Spill* (New York: Chelsea, 1976), pp. 76–85.
9. Robert S. Erikson, "The Advantage of Incumbency in Congressional Elections," *Polity* 3 (1971):395–405.
10. *Houston Post*, March 20, 1976.
11. Raw data are from the Texas Secretary of State's office; compilation by the author.
12. *Fort Worth Star-Telegram*, March 3, 1976; *Houston Post*, April 23, 1976; *Austin American-Statesman*, April 29, 1976.
13. *Austin American-Statesman*, March 4, 1976; April 2, 1976; *San Antonio Express*, March 19, 1976.
14. *Dallas News*, April 4, 1976.
15. *Austin American-Statesman*, March 20, 1976.
16. Ibid., April 23, 1976.
17. *San Antonio Express*, April 16, 1976.
18. *Fort Worth Star-Telegram*, March 11, 1976; *Austin American-Statesman*, April 9, 1976; *Dallas Times-Herald*, April 21, 1976.
19. *Fort Worth Star-Telegram*, May 7, 1976; *Houston Post*, May 23, 1976.
20. *Houston Post*, May 23, 1976.

21. *Austin American-Statesman*, May 19, 1976.
22. *Houston Post*, June 2, 1976.
23. These totals are compilations from raw data supplied by the Texas Secretary of State's office.
24. *Dallas News*, December 3, 1979.
25. *Austin American-Statesman*, May 2, 1980.
26. See, for example, the *TIPRO Reporter*, Fall 1979, pp. 6–7.
27. *Austin American-Statesman*, May 2, 1980.

9. Lessons

1. Joseph A. Schumpeter, *Capitalism, Socialism, and Democracy*, 3d ed. (New York: Harper and Row, 1960), p. 132.
2. Elting E. Morison, *Men, Machines, and Modern Times* (Cambridge, Mass.: M.I.T. University Press, 1966), p. 9.
3. Arthur Koestler, *The Act of Creation* (New York: Dell, 1964).
4. Schumpeter, *Capitalism*, pp. 134, 139–141.
5. My colleague Alfred Watkins provided me with the inspiration for the following argument.
6. Anthony Sampson, *The Seven Sisters* (New York: Bantam, 1975).
7. William A. Johnson, Richard E. Messick, Samuel VanVactor, and Frank R. Wyant, *Competition in the Oil Industry* (Washington, D.C.: George Washington University Press, 1975).
8. Alfred D. Chandler, Jr., *Strategy and Structure* (Cambridge, Mass.: M.I.T. University Press, 1962).
9. C. Wright Mills, *The Power Elite* (New York: Oxford University Press, 1959), p. 100.
10. David J. Teece and Henry Ogden Armour, "Innovation and Divestiture in the U.S. Oil Industry," in David J. Teece (ed.), *R & D in Energy* (Palo Alto: Stanford University Press, 1977), pp. 65–75.
11. The scholarly literature on this question gives some reason for thinking that there is truth in this assertion. Some investigators have concluded that large bureaucratic structures inhibit invention and that creativity is more likely to come from small or medium-sized firms in an industry. See Morton I. Kamien and Nancy L. Schwartz, "Market Structure and Innovation: A Survey," *Journal of Economic Literature* 13, no. 1 (March 1975): 9, 10; for an application of this principle to the petroleum-refining industry, see John Lawrence Enos, *Petroleum Progress and Profits: A History of Process Innovation* (Cambridge, Mass.: M.I.T. University Press, 1962), pp. 234–236.
12. John Kenneth Galbraith, *The Affluent Society* (New York: New American Library, 1958), p. 86.
13. Michel T. Halbouty, "What the Nation Needs Is More Dry Holes," *International Oil Scouts Association* 14, no. 9 (September 1973): 18–23.
14. *Newsweek*, December 24, 1979, p. 15.
15. Ruth Sheldon Knowles, *The Greatest Gamblers* (Norman: University of

Oklahoma Press, 1978), pp. 364–367.

16. *Oil and Gas Journal*, April 22, 1963, p. 169.
17. Estimates of recoverable reserves should be treated with great caution, since they rest on assumptions about economic conditions and technological advances in the future. Figures on field reserves are from Marathon Oil's exhibit no. 29, docket no. 8-67,952, Railroad Commission. (I am grateful to Warren Anderson for supplying me with this information.) Information on discovery comes from the following sources: Fancher, Whiting, and Cretsinger, *Oil Resources of Texas*, pp. 83, 112, 210, 285, 303, 310; Halbouty, *Ahead*, p. 168; Knowles, *Greatest Gamblers*, p. 296; Railroad Commission Office of Information Services.

    Of the twelve largest fields in Texas, the following were discovered by major companies: Kelly-Snyder, Levelland, Slaughter, West Hastings, and Webster. The following were discovered by independents: Conroe, East Texas, Hawkins, Seminole, Wasson, and Yates. The remaining field, Tom O'Conner, was a joint discovery.
18. Railroad Commission Oil and Gas Docket no. 126, order no. 8-16,435, September 12, 1949.
19. George Buckles, "Water Flooding in the South Ward Field," *Proceedings for Second Oil Recovery Conference*, bulletin 11 (Austin: Texas Petroleum Research Committee, 1951), p. 170; A. B. Dyes, P. H. Braun, and B. E. Coles, Jr., "Flooding in South Ward—Analysis of a Lease Performance," *Proceedings for Tenth Oil Recovery Conference* (Austin: Texas Petroleum Research Committee, 1957), p. 49.
20. Railroad Commission Oil and Gas Docket no. 126, order 8-30,250, September 27, 1954.
21. Knowles, *Greatest Gamblers*, p. 295; McKie, "Market Structure and Uncertainty," p. 552.
22. McKie, "Market Structure and Uncertainty."
23. I am indebted to David Patterson, assistant geologist of Logue and Patterson, for supplying me with this information.
24. Lovejoy and Homan, *Oil Conservation*, p. 273.
25. Adelman, "Efficiency," p. 104.
26. Once again, estimates of reserves should be treated with caution. Information on strippers is from the National Stripper Well Survey for 1977, pp. 5, 7. Information on Texas production and reserves is from the *Texas Almanac, 1978–1979*, pp. 401, 402.
27. McDonald, *Petroleum Conservation*, pp. 62–63.
28. Grant McConnell, *Private Power and American Democracy* (New York: Random House, 1966), pp. 246–297; Bernstein, *Regulating Business*, pp. 264–267.
29. *Buckley* v. *Valeo*, 424 U.S. 1 (1976).
30. Anthony Downs, *Inside Bureaucracy* (Boston: Little, Brown, 1967); Robert N. Kharasch, *The Institutional Imperative* (New York: Charterhouse, 1973).

Index